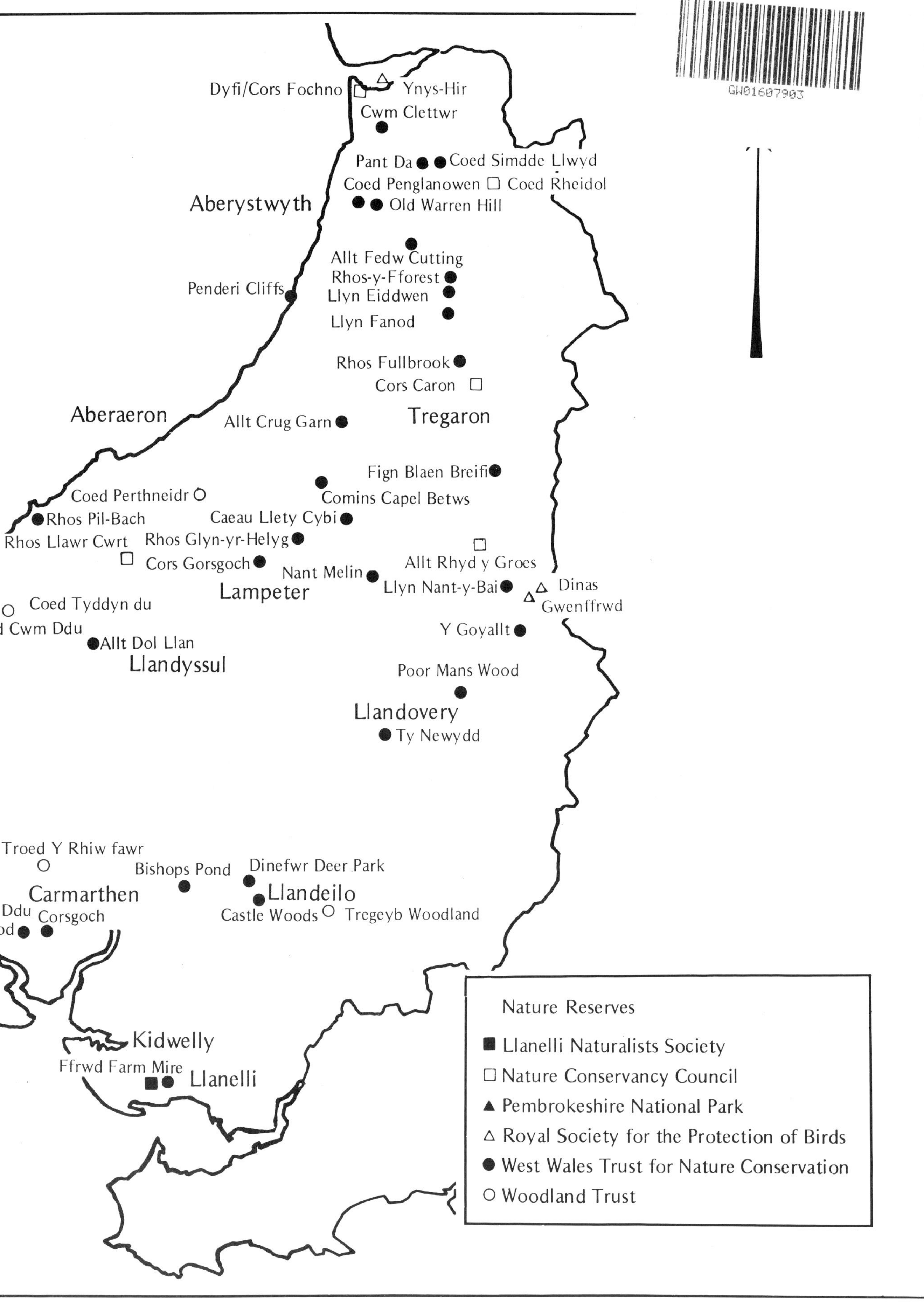

Dyfi/Cors Fochno
Ynys-Hir
Cwm Clettwr
Pant Da
Coed Simdde Llwyd
Coed Penglanowen
Coed Rheidol
Aberystwyth
Old Warren Hill
Allt Fedw Cutting
Rhos-y-Fforest
Penderi Cliffs
Llyn Eiddwen
Llyn Fanod
Rhos Fullbrook
Cors Caron
Aberaeron
Allt Crug Garn
Tregaron
Fign Blaen Breifi
Coed Perthneidr
Comins Capel Betws
Rhos Pil-Bach
Caeau Llety Cybi
Rhos Llawr Cwrt
Rhos Glyn-yr-Helyg
Cors Gorsgoch
Allt Rhyd y Groes
Nant Melin
Lampeter
Llyn Nant-y-Bai
Dinas
Coed Tyddyn du
Gwenffrwd
d Cwm Ddu
Y Goyallt
Allt Dol Llan
Llandyssul
Poor Mans Wood
Llandovery
Ty Newydd
Troed Y Rhiw fawr
Bishops Pond
Dinefwr Deer Park
Carmarthen
Llandeilo
Ddu
Corsgoch
Castle Woods
Tregeyb Woodland
od
Kidwelly
Ffrwd Farm Mire
Llanelli
Nature Reserves
Llanelli Naturalists Society
Nature Conservancy Council
Pembrokeshire National Park
Royal Society for the Protection of Birds
West Wales Trust for Nature Conservation
Woodland Trust

The Nature of West Wales 1986
has been published as a
limited edition of which this is

Number 63

a complete list of the
original subscribers is printed
at the back of the book

THE NATURE OF WEST WALES

Chough over Marloes

OVER: Puffins at Skomer, by Elisabeth Gardner.

With love and best wishes,
Liz
19.5.86

With best wishes
David Saunders
19 May 1986

THE NATURE OF WEST WALES

THE WILDLIFE AND ECOLOGY OF THE COUNTY OF DYFED

EDITED BY

DAVID SAUNDERS

with cover illustrations by

ELISABETH GARDNER

Drawings and photographs by

MEMBERS OF THE WEST WALES TRUST FOR NATURE CONSERVATION, AND THE NATURE CONSERVANCY COUNCIL, BRECON BEACONS NATIONAL PARK AND THE NATIONAL MUSEUM OF WALES

Published with the co-operation and in aid of
THE WEST WALES TRUST FOR NATURE CONSERVATION

FOREWORD BY RONALD M. LOCKLEY

BARRACUDA BOOKS LIMITED
BUCKINGHAM, ENGLAND
MCMLXXXVI

THE NATURE OF BRITAIN SERIES

PUBLISHED BY BARRACUDA BOOKS LIMITED
BUCKINGHAM, ENGLAND
AND PRINTED BY
BURGESS OF ABINGDON LIMITED
ABINGDON, ENGLAND

BOUND BY
THE CAMELOT PRESS LIMITED,
SOUTHAMPTON, ENGLAND

COLOUR PLATES AND JACKET
PRINTED BY
CHENEY & SONS LIMITED
BANBURY, OXON

LITHOGRAPHY BY
CAMERA GRAPHIC LIMITED
AMERSHAM, ENGLAND

TYPE SET IN BASKERVILLE BY
CREATIVE SERVICES
AYLESBURY, BUCKS, ENGLAND

ISBN 0 86023 242 5

Contents

List of Colour Plates

Carmarthenshire: But the principal river of the county, and one of the noblest in South Wales is the Twyi . . . The air of wildness and impetuosity, which marked its passage through its native hills, gradually softens as it advances into the more level vales . . . and changes imperceptibly into an aspect of greater grandeur and majesty.

Reverend T. Rees, 1815.

Ceredigion: Cardiganshire begins in the north in sand, pebbles and dunes and an estuary, that of the Dyfi which looks to a beautiful circle of the hills of mid-Wales.

W.M. Condry, 1970.

Pembrokeshire: Neyther perfect fquare longe nor round but fhaped with diverfe Corners, fome fharpe, fome obtufe, in fome places concave, in fome convex . . . where the wefterlye and fouth weft winds are found verye fharp and tempftuofe.

George Owen, 1603.

Puffins

Foreword

by Ronald M. Lockley

Some sixty years ago I set foot for the first time on the far western islands off the Pembrokeshire coast — to be precise on Skomer on 30 May 1926 — and experienced that mystical joy of living summer days and nights amid thousands of spectacular seabirds, which fortunately you may still enjoy today, thanks to the enterprise of our West Wales Trust for Nature Conservation.

Several species of seabirds at that time were little studied. I count myself fortunate that in the following year I realised a boyhood dream, by obtaining the lease of, and settling to live for the next dozen years upon, the more remote but not less beautiful island of Skokholm. I was happy as a sort of modern Crusoe, surviving by fishing each summer, the produce of a garden, sheep, goats, rabbits; peat and driftwood for firing, but above all studying, not only the huge colonies of seabirds, but the surprisingly numerous migrant land birds passing through.

Earlier this lands end of South Wales, today known as Dyfed from the former Welsh princedom of that name, had long drawn me, as a boy reared in industrial east Glamorganshire, to seek my holidays westwards, wandering afoot in search of wild birds and flowers; enthusiastically — even rapturously — recording these in simple diaries which I still possess.

Natural history was then my passion; but you cannot live and explore long in Dyfed without becoming profoundly aware of its history of human settlement. As a young man I vividly remember exploring limestone caves on Caldey Island and near Tenby, at that time the haunt of large colonies of horse-shoe bats, and examining the rich collection of prehistoric animal bones. These sites were said to be the lairs and larders of the first Man of Knowing, *Homo sapiens,* who vanished with the last glacial period. Fauna and flora, geology, human ecology, all are part of the grand design comprising *The Nature of West Wales*. This book introduces all to us, with the newest interpretation of the past, and of our present knowledge of the natural sciences of Dyfed.

My diaries of the late twenties and early thirties speak of far from idyllic times as to the conservation of nature. They tell of such abuses as visitors digging up and carrying off rare plants. Fishermen persecuted seals as destructive of fish and nets; they were shot freely, helped at times by trigger-happy 'sportsmen' for the fun of it, or for a trophy head! Peregrine eyries were robbed of eyases for falconry. Collecting the eggs of sea-birds, choughs, ravens, buzzards and other scarce species was rife. The Royal Society for the Protection of Birds in London was first in the field to try to promote a system of local volunteer and part-paid wardens, in an attempt to protect rare species, especially cliff nesting birds. Some of the part-time men so employed I knew to be undercover agents for falconers, collectors, and so-called sportsmen. The situation did not improve until, with the support of the RSPB, the Pembrokeshire Bird Protection Society was inaugurated early in 1938, Lord Merthyr (President), Lionel Whitehead (the new owner of Ramsey Island) as Chairman, and myself as general dogsbody secretary. Unfortunately Whitehead died suddenly within that year, and soon the war began. We struggled on, fighting a Government demand that all the peregrines must be shot, because these falcons were intercepting trained homing pigeons carrying vital war communications. In Dyfed the falconer Jack Howell was appointed marksman, but I can now say that, as the observer appointed to monitor his shooting, between us we spared many a fine adult, and handreared in secret places as many of the eyases we could lift from Dyfed eyries.

I need add no more here. In a sense, the diversion of fishermen, sportsmen and collectors to their obligations elsewhere during the war was largely a helpful interregnum in the war on nature. The war over, the Society blossomed into the full scale study and conservation of all aspects of wild nature as described elsewhere in this volume. As a founding father (the editor's words in inviting me to write this foreword) I feel honoured to do so, delighted with this new updated interpretation and description of the work of the West Wales Trust for Nature Conservation. This book fulfils a considerable need and public demand for information on the true Nature of West Wales.

New Zealand
February 1986

Otter: West Wales is a key area for this scarce mammal. (HTC)

Introduction

In 1938 the Pembrokeshire Bird Protection Society was founded to become after several changes of name the present day West Wales Trust for Nature Conservation. It is appropriate that within less than two years of its Golden Jubilee the Trust should present this book *The Nature of West Wales*. Where is West Wales you may ask; is Aberystwyth there or Ammanford? For the purposes of the Trust, and for the purposes of this book, West Wales embraces the whole of the county of Dyfed — the vice-counties of Carmarthen, Ceredigion and Pembrokeshire. From the marches of Powys and West Glamorgan westwards to St George's and the Bristol Channel lies Dyfed, a massive and varied county some 6,090 square kilometres in extent.

The West Wales Trust for Nature Conservation has for nearly 50 years been the only voluntary nature conservation body devoted to the whole of Dyfed. It has striven to achieve its aims in the face of increasing threats to the countryside and its wild inhabitants, with extremely slender resources. One hopes that our founding fathers, those who are still with us, and those who now catalogue and enjoy the myriad of exotic species in celestial woods and marshes, will appreciate the efforts of their successors. It is with particular pleasure that a founder member of the Trust, who subsequently urged us forward for so many years, R.M. Lockley, should write the Foreword to this book.

A great deal has been written about the natural history of West Wales, but much is hidden away in journals and other works and needs to be teased out. It is hoped that *The Nature of West Wales* will bring some of this information to the attention of the general naturalist; for those who require more, it will point the way along the sometimes tortuous trackways of the researcher. The book is meant to be a guide, a stimulus, and to provide enlightenment to all those who enjoy, and wish to know more about the wildlife of the county.

I am extremely conscious of the debt which the Trust owes to the many contributors; without their generous assistance the book would not have been possible and the Trust gratefully acknowledges their considerable help. The artists and photographers also deserve a special mention, their work helping immensely to bring the beauties and interest of the wildlife of West Wales to the reader. Each is acknowledged in the captions, to which there is a key immediately following the references in the appendices. A special mention must be made of Mrs Elisabeth Gardner, who not only provided the colour illustrations depicting scenes from the nature reserves of Skomer and Nant Melin, but also provided the chapter headings and some other illustrations.

Thanks are also due to Miss Sian Miller and Miss Nicola Wilkins, who undertook the typing of preliminary drafts, those in my handwriting being especially troublesome, and to Mrs Jackie Lawrence for typing the final manuscript. I am indebted also to my colleagues, Mrs Eileen Williams and Miss Brenda Hibberd, who have together shared much extra work during my frequent absences from the Trust Office during the final stages of the book. Special thanks are also due to Miss Elinor Gwyn for assistance with the Welsh names included in the species index.

The Trust owes much to its publisher, Clive Birch of Barracuda Books, both for his patience when the manuscript was delayed, and for the prompt way that production schedules were re-arranged as soon as it arrived.

Lastly, our thanks go to all our readers, those who kindly placed advance orders, and those who have purchased the book since. All royalties from sales go to the West Wales Trust for Nature Conservation to further its work in the county of Dyfed.

DAVID SAUNDERS
Haverfordwest, Dyfed
January 1986

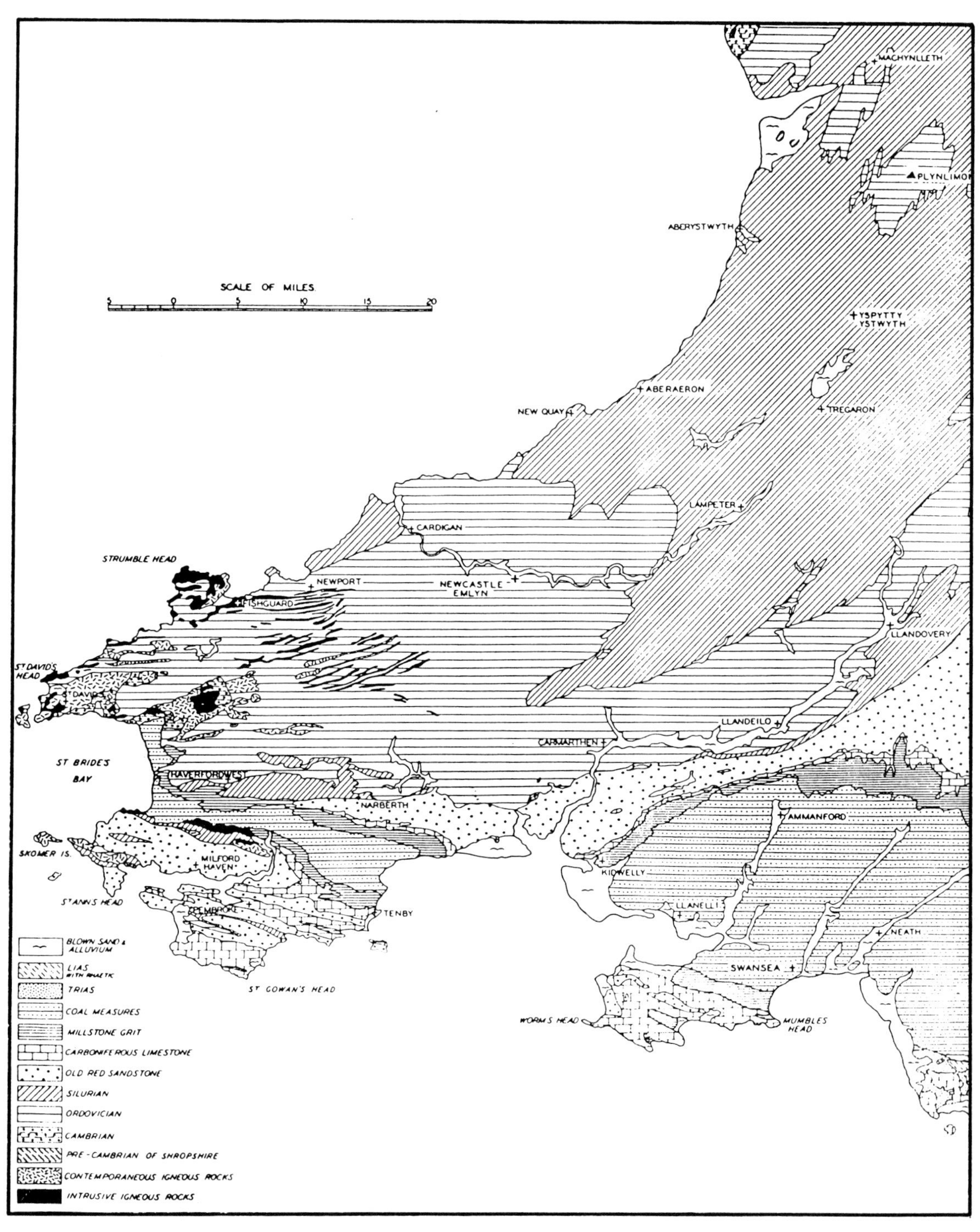

Simplified outline map of the geology of South Wales (adapted from *British Regional Geology, South Wales,* published by HMSO).

Origins and Influences

Treffgarne Rocks

Dyfed was the name given to the amalgam of the three former counties of Cardiganshire, now referred to as Ceredigion, Carmarthenshire and Pembrokeshire, at the reorganisation of local government boundaries in 1974. These three counties had remained separate entities since their formation under the Act of Union in the reign of Henry VIII. The new name, however, also has its roots in history for, in former times, it had been given to this south-western peninsula of the then Kingdom of Deheubarth.

Man has been active here since earliest times. Massive stone tombs, the burial places of tribal chiefs and priests, like that at Pentre Ifan in North Pembrokeshire, may still be seen. There are the remains of stone circles and alignments, while standing stones, ghosts from man's beginnings, dot the landscape for us to ponder over. From rocky tors on Mynydd Preseli huge 'bluestone' boulders were dragged to build the inner ring of Stonehenge, their transportation surely one of the wonders of the ancient world. The massive promontory and hill forts of Iron Age man, like that at Pen Dinas, Aberystwyth, are everywhere. Soon afterwards the Romans came this way and there is much evidence of their occupancy especially in the east, where the gold mines of Dolaucothi, and their road from Llandovery northwards, Sarn Helen, are reminders of their presence. Place names on the coast remind us that the Vikings harried here, and in geological time that brings us right up to the present and the landscape and character of the county today.

Geographically Dyfed is very much a representative Welsh county. There is a ruggedness in its scenery in the western Black Mountains or Mynydd Du in the south and in the Plynlimon range in the north, that is similar to the mountain ranges of mid and north Wales. The plateau lands of its middle and western regions are typical of vast tracts of such land in the Principality, while the rich arable and pastureland of the coastal strip and broad river valleys typifies much of the lowlands.

In order to appreciate fully the strong formative influences which have structured the present landscape of Dyfed, we need to go back millions of years. Although a cursory glance at its surface rock structure may give the impression of dull uniformity, this is not really the case. Vast areas may be dominated by rocks of a particular era; nevertheless there are pockets where the majority of the British foundational types occur. Over most of its land mass Dyfed is underpinned by Pre-Cambrian and older rocks, but the ones which determine its modern structure are superimposed above this basal foundation. The slower weathering rocks of the Cambrian and Ordovician eras are exposed in the southern region, while the dominant rock of the north is the softer Silurian. It is the friability and easy weathering characteristics of these Silurian shales and mudstones which give rise to the gentle rounded topography of much of the region, and which are such a feature of the Cambrian Mountains, the backbone of eastern Dyfed, and of its westward splay towards the Preseli range.

It is in the southern region of the county that we have the most varied geological patterning. Younger rocks than those of the Ordovician series occur towards Carmarthen Bay and westwards. These are the Carboniferous Limestones and Coal Measures. Towards their extremity in Pembrokeshire they have interspersed within them, especially in the coastal regions, volcanic igneous rocks. Indeed, it is these outcrops that are often the dominant feature of the landscape and are readily seen in the Fishguard hinterland. Where the soft and hard rocks occur together and intermingle on the coast, their differing weathering properties give rise to an interesting coastline of rugged cliffs and sandy shores. A further feature of the southern coastal belt is the distinctive rich reddish brown soil overlaying the Old Red Sandstone of the Devonian era. This foundational rock type is the dominant feature of the English south-west Midlands, but it tongues through as an obvious originator of the surface soil in south east Dyfed, before plunging beneath the sea of Carmarthen Bay.

Like many other British counties, Dyfed derives much of its wildlife interest from its position relative to the sea, a fact which also accounts for its maritime climate of high rainfall and mild winters. It has a coastline extending to some 400 kilometres and a large part of the county is within less than 30 kilometres of the sea or estuarine waters. Its shape reminds us of that small creature of our earliest studies in the natural world, the amoeba, for it has a nondescript body, thin towards the north, a rather fleshy south, but with a broad outstretched pseudopodium giving the appearance of movement towards the south-west. The land surface extends to 6,090 square kilometres; its maximum length from north-east to south-west is 150 kilometres.

The topography of Dyfed determines in large measure its climate, although the near flow of the Gulf Stream has also a valuable role to play. The land standing about 60m above sea level rises abruptly on the western seaboard to two higher plateaux ridges at 180m - 200m and above the 500m contour. Moisture laden sea air carried prevailingly by south-westerly winds has to ascend rapidly on reaching land. This has the consequence of producing a coastal strip narrow in the north but quite extensive in the south, which is drier, warmer and sunnier than the uplands. Coastal rainfall averages 1,000mm annually, but this figure rises rapidly to 1,250mm within relatively short distances inland. It can exceed 1,500mm on the peaks of Plynlimon, Mynydd Du and Mynydd Preseli. The cloud cover associated with this rainfall depresses air and soil temperatures quite markedly inland and besides, reduces the number of sunshine hours received. The extent of cloud influence is not always appreciated, but the heads of river valleys can have 1.5 hours less bright sunlight daily throughout the year than coastal regions.

Although these statistics are in themselves of considerable interest to the nature lover, it is the summation of all climatic factors, including day length and exposure to wind, which is reflected in flowering times and other aspects of growth and development in both plants and animals. A convenient way of seeing the magnitude of locational differences is to examine floral isophenes. These are constructed from first flowering data of some ten to fifteen selected species from a range of sites within a region.

When such data is collected annually over a number of seasons the sites of equal average earliness can be mapped. It is noticeable that there is a marked delay in flowering as one ascends higher ground. Whereas the mean date of flowering for the recorded species occurred on or before 27 April in western Dyfed, it was up to three weeks later in the inland mountain areas.

The traditional farming of the area has clearly been influenced by climatic factors. For instance, Pembrokeshire has been noted for its early potato growing, and also sustained a small flax industry during the 1939-45 war. Studies at Aberystwyth for the post-war period showed that there was a geographic basis for the calving pattern of Dyfed's extensive dairy herds, which again could be linked strongly to climatic regional differences. The gross influence of climate and altitude on the distribution of native grasslands has also been demonstrated. The poorer grasslands of the hills consist of those species which are adapted not only to acidic conditions —itself a factor having a strong climatic component — but are also species that can endure summer exposure to lower temperatures and reduced insolation. Thus the upland rough grasslands are dominated by mat-grass, purple moor-grass and sheep's fescue, whereas in the less rigorous climate of lowland area bents, Yorkshire fog, creeping soft-grass, perennial rye-grass and cock's foot provide the grazing for stock.

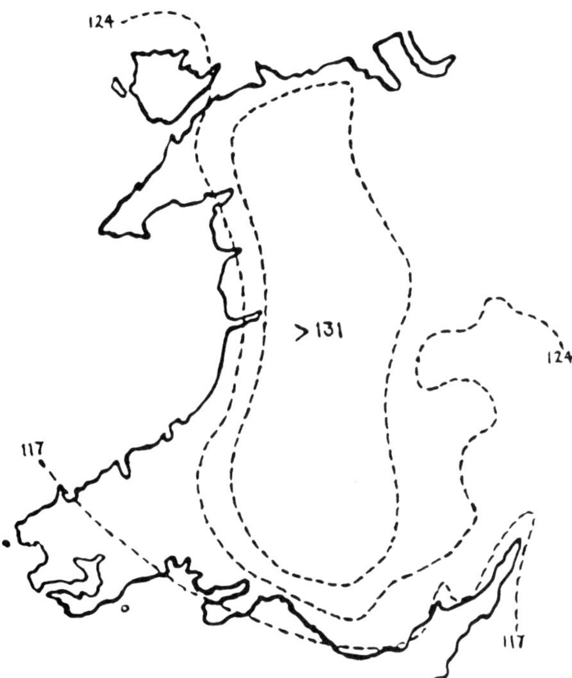

Average floral isophenes (from Bowen, E.G. *Nature in Wales* 1: 17-21).

The upland regions have been used increasingly for grazing, especially since the war years. It had been a tradition for these sheep walks to be grazed until the weather became intolerable, when the ewes and yearling sheep were brought down to the lowlands, the hendre, before returning again when there was sufficient grazing in the spring to their summer grazing on the hills, the hafod. The severity of the summer grazing accounts to a large extent for the barren appearance of the hills, and recent intensification of sheep production at the expense of cattle, which also used to roam the hills in greater numbers than at present, has resulted in a rapid spread of bracken; the area which it now infests has doubled in thirty years. Such spectacular responses illustrate the dynamic nature of change and show how seemingly trivial alterations in management can drastically alter vegetational patterns. Indeed, the vast area of heather moorland which used to dominate much of the hill land has largely disappeared.

The most interesting ecological areas are those that are not so dependent on man's interference but have a degree of natural protection from the ravages of overgrazing. The gorges, deeply cut river valleys and gullies, which occur along major geological faults, are typical of the upland scene. They remain havens of floristically rich wooded slopes, often covered by mosses and lichens whose luxuriant growth give adequate testimony to the shelter and high humidity that these clefts in the hills provide for them.

Few natural lakes occur within the county borders, but the increasing demands for water have meant large reservoirs being created in the upper catchment areas of some of Dyfed's rivers. Llyn Brianne in the upper Tywi covers 88 hectares and impounds 66 million cubic metres. Llys-y-Fran, south of Mynydd Preseli, has a capacity of 21-25 million cubic metres, while the largest man-made power generating system in Wales at Nant-y-Moch, in the Plynlimon range, holds 26 million cubic metres at full capacity.

Aspect is another important consideration of the plateau and glaciated river valley environment. It produces an intriguing variation in what might, at first sight, appear to be a rather dull and colourless habitat. It produces nuances of light and shade, exposed and sheltered, moist and dry, steep sided and gentle slopes that add a wonderful richness to the vegetation and animal wildlife within narrow confines.

Viewing the mountain plateaux from the air reveals a considerable degree of division of the land mass, and it soon becomes apparent that the river system in its entirety gives a high ratio of stream distance to unit area of land. The rivers themselves dissect the coast at a frequency of about one per 30 kilometres which, for a county of its size, shows how important a contribution rivers make in structuring its landscape. Not only do they afford typical upland habitats of great value, but their course to the sea provides a richness to the landscape and a bounty to wildlife of outstanding value. They flow in glaciated valleys which are generally broad in their lower reaches, allowing the meandering streams to cut new courses when the old ones become silted. Over a period of time records of these changed courses produce an interesting pattern, and emphasise the admixture of soil and sediment which provides the basis for vegetative growth in these river meadows.

The Smalls lighthouse

Estuaries are a particularly fine habitat and the Dyfed coast has some notable examples, from the salt marshes and mud flats of the Dyfi in the north, to the rich river marshland of the Tywi and Lougher in the south-east. Interspersed with these estuaries there are fine examples of cliffs and sand dunes. Some of these dunes are of recent origin, whereas others have been developed over the centuries. In fact the dune system is perhaps one of the best examples of the rapid changes which can occur in the land environment requiring special adaptions by its inhabitants.

The Pembrokeshire Coast National Park is the most extensive of the two designated National Parks in Dyfed and comprises 625 square kilometres of sea coast, estuarine and upland habitats. Only part of the extensive Brecon Beacons National Park is found within Dyfed, but even so its wild, mountainous terrain rising to 633m and extending to some 242 square kilometres complements its larger partner, and between them they represent the environmental extremes of south Dyfed.

The population of 326,300 is unevenly distributed throughout the county, with most of the larger towns situated in the south or on the coast. Since man has the means either to stress or structure his environment, the present quality of the region as a wildlife habitat has, to a large degree, been moulded by his activities. No visitor to South Wales, including that part of Dyfed contiguous with West Glamorgan, cannot but be startled by the ugly black conical spoilheaps left as seemingly indelible blots on the landscape. Fortunately such unwelcome products of Dyfed's heritage are localised, and the county remains largely untouched by heavy industry.

At the beginning of the 20th century industrial problems began to be appreciated. Pollution from some heavy metal processing, in the form of barren spoil heaps and smoke emissions, marred the landscape and affected vegetation in many parts of South Wales. Dyfed itself did not escape the problem. For instance, in north Dyfed, the rejuvenation between 1840-70 of the centuries' old lead, silver and zinc mine workings led to the poisoning of rivers and lakes. The heavy metal pollution of these waters persisted for a century and was only overcome by expensive purification measures. The monitoring of water quality was an important aspect of river management in this area, and indeed continues to be so.

The welfare of the countryside has in this present century been closely linked with the relative prosperity of the people. In a rural county such as Dyfed it is the prosperity of the land that has the more immediate and lasting influence. From 1927 there was a marked depression in farmers' fortunes, and land that had been in cultivation again fell into disuse. While such disused land is thought of in economic terms as derelict, its new-found role probably afforded a respite for some plant and animal species that had become rare. In a survey of the grasslands of Britain carried out in the late 1930s there were considerable acreages of poor pasture. Perhaps it was the realisation of this position, coupled with the need for home food production during the Second World War, that set in motion quite a revolution in its own way in the countryside. Unfortunately agricultural prosperity was inextricably bound with the use of fossil fuels and high fertiliser inputs. Newer chemicals were accepted at their face value and used to promote ever higher outputs of grass, grain and grazing stock. Prosperous agriculture became an end in itself, and few farmers were willing to face the consequences of maintaining high interest in the environment for its own sake. As the result of excessive productivity promoted by all governments in the post-Second World War period, the farmers of Dyfed, as in the whole of the western world, are again faced with the need to cut back production. Once more there is a cry that the environment *per se* is important for the community, so that at last we may be awakening to a new vision.

Let us hope that our county will, with the new goodwill of all landowners and country dwellers, continue to be a haven for the wildlife of which we are justifiably proud. We invite you to explore some of the habitats of Dyfed in this book, and trust that the glimpse you will have of Dyfed's rich heritage will create a desire for a fuller investigation.

Pentre Ifan

ABOVE: Iron Age ditch and rampart, Castell Mawr, Crosswell. (SL)
BELOW: Mynydd Preseli from the north. (SL)

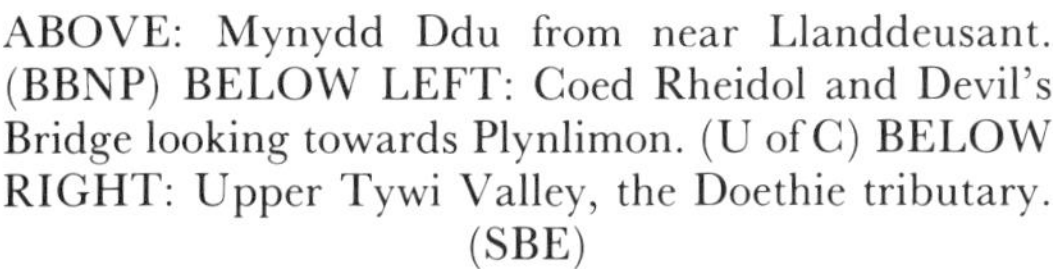

ABOVE: Mynydd Ddu from near Llanddeusant. (BBNP) BELOW LEFT: Coed Rheidol and Devil's Bridge looking towards Plynlimon. (U of C) BELOW RIGHT: Upper Tywi Valley, the Doethie tributary. (SBE)

West Wales Uplands

Uplands

Merlin

'Upland' is defined here as the open land above the upper limits of traditional field enclosure. In Wales the traditional zone between upland and lowland, where the highest enclosures are found, is known as the *ffridd*. The *ffridd* and the area just above it are often characterised by bracken, gorse, or grassland with patches of hawthorn scrub. At higher altitudes, upland heath or blanket bog often become dominant.

In Dyfed, much of the upland is over 400m in altitude, the highest summits being Plynlimon at 752m and Bannau Sir Gaer on Mynydd Du at 750m. In places however, the unenclosed land extends down to below 300m and well below this level in the hills of Pembrokeshire.

The map shows the main upland areas in West Wales. The uplands between Plynlimon and Mynydd Mallaen take the form of a deeply dissected plateau, inclining towards the south. The rocks are Silurian and Ordovician slates and shales, which give rise to rather acidic nutrient-poor soils, except in a few localities where the presence of more base-rich beds of the Llandovery series allows a richer soil to develop. Mynydd Du and Mynydd Preseli are quite different in character. The great north and east-facing scarp of Mynydd Du is formed of Old Red Sandstone of Devonian age, in which base-rich cornstone beds are present. A narrow band of Carboniferous Limestone outcrops along the northern edge of Mynydd Du. Both cornstones and limestone produce base-rich soils which support a wide variety of plants. The southern part of Mynydd Du is composed mainly of acidic Millstone Grit, and the vegetation there is similar to that on the slates of the hills further north. Mynydd Preseli is formed of Ordovician rocks capped by tors of intrusive igneous rock. Some of the volcanic rocks are base-rich, and this is reflected in the range of plants found on the lower slopes, where the soils are influenced by the base-rich drainage water from the higher outcrops.

All the West Wales uplands have been grazed for many centuries. For over two hundred years the pastoral economy has been based upon sheep, with relatively small numbers of ponies and beef cattle. Before that, there were many more cattle on the hills in summer. Wild deer disappeared in medieval times. The open uplands owe their present treeless, mostly grassy, character to this grazing regime, and the associated burning of rank vegetation. There is ample evidence from place names and from the preserved remains in peat bogs that the hills were formerly covered with scrub woodland, chiefly of birch and willow above about 400m, and sessile oak below. Very little of this native forest remains.

Stocking densities of sheep on the Dyfed uplands are generally higher than those on the hills of similar altitude elsewhere in Britain. The West Wales hills have relatively productive soils, and the climate tends to be milder. Breeding ewe numbers on the sheepwalks average around 1.5 per hectare and sometimes reach 2.5 per hectare (one per acre). In Ceredigion, and locally elsewhere, entire breeding flocks may remain on the hills throughout the year, only the lambs and draft ewes being removed in autumn. In other districts some or most ewes are removed to lower ground before the end of the year, and lamb there before they are returned to the hills in spring. There are many permutations between the two extremes, depending particularly upon the availability of lowland winter grazing; the pattern is changing with the enclosure and improvement of marginal hill land and the introduction of lambing sheds, to enhance productivity. Sheep are now much more valuable than formerly, so there is a trend towards better shepherding and the provision of supplementary feed on the open hill, to reduce mortality and improve the quality of the end product. Grazing pressure upon the natural vegetation is tending to increase, as is the economic impulse towards agricultural improvement. Much of the hill land in Dyfed is potentially improvable so, given a favourable economic and political climate (and taking into account the provision of grant aid), there must be a question mark over the continued existence of some of the remaining sheepwalks and moorlands as we know them today. At present there are two 'non-economic' constraints which resist change. One is the existence of large areas of common grazings, and the future management of these commons is currently under review; the other is the legal provisions of the Wildlife and Countryside Acts, which include the notification of Sites of Special Scientific Interest, and the protection conferred upon certain moors and heaths within National Parks. Often these various categories of land coincide or overlap.

There have been other changes in the Dyfed uplands in the past half century, more radical in effect than those produced by agriculture. The construction of new roads and reservoirs brought many more people into remote areas for recreation, and modified existing habitats. The hills are no longer the quiet remote places they were. The uplands, especially Plynlimon and Elenydd, suffer gross noise and exhaust pollution from the constant exercises by low-flying military aircraft. Acidification of bogs, pools, and watercourses, caused by the precipitation of airborne combustion pollutants from distant industries upon the hills, is thought to be affecting animal and plant life. The most extensive physical change by far has been afforestation with exotic conifers; Sitka spruce, hybrid larch, and lodgepole pine are the chief species grown in upland situations. Afforestation in upland West Wales commenced between the two world wars, but expanded most rapidly during the fifties and sixties, when sheep farming was depressed and hill land prices were low, and when tax concessions also favoured private planting. Little new planting was done after the early seventies, when sheep rearing became more profitable. In North Carmarthenshire and Ceredigion, where much of the upland was privately owned, over a third of the area which was open sheepwalk in the thirties is now forest. Further south, the high proportion of common land limited the extent of planting.

Afforestation removes the natural plant and animal communities of the hills, except for small areas left unplanted for various reasons, or managed deliberately to favour amenity or wildlife conservation. It can also have an impact upon the wildlife of adjacent open land, by harbouring predatory species, modifying drainage or local climate, and improving human access. On the other hand, a whole new range of animals will appear in areas which may previously have held little variety; a few existing hill species are favoured by the change. The main need is to try to ensure that the open areas most vital to the survival of viable upland plant and animal communities remain unmodified. For hill birds particularly this involves large tracts of land, because they need big breeding territories or extensive hunting grounds.

Scattered alpine clubmoss and fir clubmoss occur among the sheep's fescue and mat grass on the summit of Plynlimon. These clubmosses are widespread on the upper slopes of the mountains of Snowdonia, but Plynlimon is the most southerly Welsh mountain on which they occur among the summit grasslands in any quantity. Further south in Wales they are confined to a few high rock ledges and crags. The dwarf willow also occurs on the summit ridge of Plynlimon near its southern limit in Britain; its short woody stems and rounded leaves can be hard to find among the grazed bilberry plants nearby.

North of the summit ridge, on shallow soils on steep slopes, are large areas dominated by heather, with scattered crowberry and cowberry. On flatter ground this grades into blanket bog on deep peat, dominated by heather and tussocky cotton grass. In the wetter areas *Sphagna* (bog mosses) are abundant, with round-leaved sundew and the trailing stems of cranberry. Blanket bog and heath dominated by heather were once more widespread in mid-Wales, but the dwarf shrubs have been lost from many of these areas as a result of prolonged periods of heavy grazing, sometimes combined with burning. Most of the lower slopes of Plynlimon now carry acidic grassland or blanket bog composed of tussocky purple moorgrass.

Many of the more unusual plants found in the uplands are associated with flushes, that is areas where water movement through the soil carries a supply of nutrients and prevents stagnation. Starry saxifrage, a common constituent of wet mossy flushes in the mountains of North Wales, occurs on the steep north-facing slope of Plynlimon above the corrie lake Llyn Llygad Rheidol, at the southern limit of its range. Other interesting plants found here include mossy saxifrage and viviparous fescue. The latter reproduces by producing small plantlets rather than seeds, whose production may be limited by the low temperatures and lack of sunshine at high altitudes.

The central block of open land around the summits of Plynlimon and Mynydd Mallaen consist of rounded hills and ridges with vast areas of acidic grassland and blanket bog, much of which is dominated by tussocks of purple moor-grass. A few relatively small areas of heather and cotton-grass survive, mostly along the higher ridges, and the uncommon bog rosemary and the mud sedge may occasionally be found in these. Gors Lwyd, on the col between the Ystwyth and the Elan, is the most notable area of mire, with abundant *Sphagna* and a well-defined system of small pools and hummocks of tussock-forming vegetation. The West Wales Trust for Nature Conservation (WWTNC) reserve at Figyn Blaen Brefi has a similarly well-defined pool and hummock system, though on a smaller scale. This structure is characteristic of wet, relatively undisturbed blanket mire, and is now uncommon in West Wales, due to the effects of heavy grazing and burning. These hills are dissected by steep rocky stream valleys, often containing sessile oak woods, of which a fine example is provided by Allt Rhyd-y-groes, a National Nature Reserve in the Doethie valley. Otherwise variation is provided mainly by small localised flushes, in which ivy-leaved bell-flower, whorled carraway, and occasionally the white beak sedge, may be found. In one locality in the Camddwr valley the least clubmoss occurs, at its most southerly station in Wales.

Many small lakes and pools are present among these hills, particularly in the area north and east of Strata Florida. These are typically oligotrophic, nutrient-poor, and contain a limited but characteristic flora including water lobelia, narrow-leaved bur reed, bog pondweed, and shoreweed. Disused lead mines are found locally in these hills. Few plants can grow in the lead spoil, but one or two, including the rare forked spleenwort, can tolerate soils rich in heavy metals.

Much of the remaining open hill land in Elenydd and the Cwm Doethie/Mynydd Mallaen uplands is included within Sites of Special Scientific Interest. There are several large areas of common grazings. The Old Red Sandstone cliffs of Bannau Sir Gaer and Bannau Brycheiniog make up the dramatic scarp face of Mynydd Du which overlooks the corrie lakes Llyn-y-Fan Fach and Llyn-y-Fan Fawr (the latter outside Dyfed). The wet ungrazed cliff ledges support a rich flora, including such species as roseroot, lesser meadow rue and ox-eye daisy, together with sea campion, which is more familiar on coastal cliffs. Mynydd Du also has a number of limestone crags and small limestone pavements. A variety of basiphilous plants such as green spleenwort, brittle bladder fern, and hairy rock cress are found on the ungrazed ledges on the limestone crags. The surfaces of the limestone pavements, the clints, are exposed to grazing and to extremes of temperature and drought, and are bare of vegetation; but in the deeper more inaccessible fissures or grikes, plants such as lily-of-the-valley and wood anemone may be seen. These are more familiar as woodland species, and favour the humid shaded conditons of the grikes. Now growing many miles from the nearest suitable seed sources, the flora of these isolated pavements may be relics of a time when broad-leaved woodland or scrub cover was extensive over all but the highest mountain areas of Wales. Another interesting plant which may be seen in grassland in the vicinity of the limestone crags and pavements is the soft-leaved sedge, a local plant in Wales.

Most of Mynydd Du is within a Site of Special Scientific Interest, and it is entirely within the Brecon Beacons National Park. It is common grazing land.

The adjacent hills of Mynydd Preseli and Carningli Common in Pembrokeshire have a unique character among the uplands of West Wales. Both are crowned by dramatic rocky tors. Carningli, the more westerly of the two, has large areas of oceanic heath dominated by western gorse, heather and bell-heather, which make a blaze of yellow and purple in late summer. On Mynydd Preseli the gorse and bell heather are largely absent, although some large areas of heather are present. On the high ground, scattered plants of fir clubmoss and stag's horn clubmoss may be found locally. One of the most interesting features of both hills is the large complex of wet heath and flushes on the lower slopes, dissected by many small streams and rills carrying base-rich seepages from the dolerite outcrops above. A number of uncommon plants are found here; they include the insectivorous pale butterwort, the lowly marsh clubmoss and the elusive bog orchid. The two areas form a Site of Special Scientific Interest and are almost entirely grazing commons. The uplands of Mynydd Llanllwni and Mynydd Llanybydder, with the adjacent but smaller Mynydd Llanfihangel Rhosycorn, have remained unenclosed and unimproved largely because they are grazing commons. Geologically and floristically they resemble Mynydd Mallaen and the southern part of Elenydd. There are extensive tracts of purple moor-grass, but this is cropped short and there is much cross-leaved heath among it. Gorse and bracken cover large areas on the western slopes. About half the site is still more or less dominated by heather, especially on Mynydd Llanllwni, but it is mostly heavily burned and grazed. No detailed vegetational survey has yet been made.

The Silurian outlier of Mynydd Bach on the Cardiganshire coastal plateau also partly avoided enclosure and improvement because of its status as common land. Most of the area is covered by fairly dry heathland-grassland communities with a good deal of heather, but there are two actively growing basin mires, Cors Pwll-yr-ych at the northern end of the site and another adjacent to the south end of Llyn Eiddwen. The lake itself, which is a WWTNC reserve, is noted for its remarkable flora. There are abundant stands of water lobelia along with carpets of shoreweed and quillwort, and the lake is one of the most southerly sites known in Britain for awlwort, which forms an unusual association with floating water plantain. Most of the area is a Site of Special Scientific Interest.

Almost all the characteristic breeding birds of the Welsh uplands occur in Dyfed, though some are uncommon or irregular. Pride of place must go to the carrion feeders, a group of birds dependent to some degree upon the large supply of sheep and lamb carrion available on the open hills, particularly in late winter and early spring. The magnificent red kite found its last British refuge in Carmarthenshire and Ceredigion early this century, and it is still there, though now spreading further afield. The recorded density of ravens, particularly in Elenydd, is higher than has been found in any other part of the wide holarctic range of the species. Buzzards are also numerous, reaching the highest densities reported in Britain. These birds, and other members of the crow family, tend to breed and roost in the valleys and scavenge over the open sheepwalks. The raptors also hunt live prey there in easier times.

Crag-breeding peregrines have now reoccupied all the larger upland areas, an astonishing recovery from near-extinction due to pesticide contamination in the sixties. Kestrels are ubiquitous, but the merlin now has only a toehold in Plynlimon and Elenydd-Mallaen and has not been reported breeding in Mynydd Du or Mynydd Preseli for some years. The hen harrier bred once in recent times in Elenydd.

Red grouse still have patchy distribution in the limited areas of suitable heather in the northern hills, south to Mallaen, and may still persist on Mynydd Llanllwni, though this is now an isolated and endangered population. On Mynydd Du they appear only irregularly probably as wanderers from Fforest Fawr to the east. The Mynydd Preseli population died out a century ago. Black grouse increased with afforestation and occur mostly near the margins of plantations, from Plynlimon south to the hills of the upper Tywi catchment, with a few recent records in Mynydd Preseli.

Apart from curlews, which are in all the upland areas, though mostly on the fringes, and a few snipe in the wetter parts of most uplands, the main hill wader populations are in the north. Elenydd is the stronghold for golden plover and dunlin, though unfortunately relatively few in the Dyfed part, but a few golden plovers also breed on Plynlimon and in Doethie-Mallaen. They have bred occasionally in the past on Mynydd Du, but not in the Pembrokeshire hills. Dunlin probably still nest in the Doethie uplands. A few redshank breed in hill bogs in Elenydd. Common sandpipers nest by lakes and streams from Plynlimon south to Mynydd Du, but are usually absent from Mynydd Preseli. A few lapwings nest on the open uplands, but more in the *ffridd* and below.

Colonies of black-headed gulls are confined to the Ceredigion hills, as are breeding teal; both species frequent the lakes and bogs. Other hill birds with restricted ranges locally include the stonechat, which breeds most consistently in the Pembroke hills, though sporadically elsewhere, and the ring ouzel, a regular but sparse breeder from Plynlimon to Mallaen and on Mynydd Du, occasional on Mynydd Preseli.

There remains a fairly long list of upland species, mostly small passerines, which are found in suitable habitat on all the West Wales hills. The commonest bird by far is the meadow pipit, often parasitised by cuckoos. Skylarks are legion on the grassy plateaux, wheatears on stony ground with short turf are remarkably common on Mynydd Du, while whinchats favour long vegetation like rushes and bracken. Grey and pied wagtails are found especially near water, dippers in most streams, wrens in crags and gullies, reed buntings in rushy bogs, linnets in gorse scrub on the hill margins. Wherever trees or bushes occur, some of the common scrub birds like chaffinch, tree pipit, or redstart may be found. After the breeding season, many other lowland birds forage on the open uplands, and flocks of lapwings, rooks, mistle thrushes, or starlings may be seen. In winter various wildfowl visit the hill lakes and bogs in the north, including whooper swans and occasional parties of Greenland whitefronted geese. Nearly all the small birds leave the hills in winter.

Hare

ABOVE: Moorland in the Gwenffrwd Reserve, Upper Tywi Valley. (TP)
BELOW: Rocky tors on Mynydd Preseli. (SBE)

ABOVE: The mountain pansy has already vanished from many upland pastures. (NCC) BELOW LEFT: Cotton grass, a dominant plant on the upland blanket bogs. (NCC) RIGHT: Common butterwort, one of the insectivorous plants of the blanket bogs and *Sphagnum* mires. (NCC)

Three waders which breed on the West Wales uplands. ABOVE: Curlew, widely distributed, but probably declining as a breeding species in West Wales. (MA) BELOW LEFT: The golden plover breeds in small numbers mainly on the high ground of the north-east. (HP) RIGHT: The snipe rarely breeds in Pembrokeshire but occurs fairly widely elsewhere in the uplands. (HP)

ABOVE: Black grouse, now almost confined to a few parts of Ceredigion. (HP) BELOW LEFT: The wheatear is widespread as a breeding bird on the uplands. (HP) RIGHT: Raven at nest. (NM of W)

The upper Rheidol Valley, sessile oak and introduced conifers. (NCC)

Woodlands

Woodlands by night

Only about 8% of the land surface of Great Britain is now wooded, while the figure for broad-leaved woodland is less and in Wales only about 3%. Yet at one time broad-leaved woodland covered more than half the land area, with only high exposed ground and the great marshes being without trees. From Mesolithic times man has been successively removing woodland. At first this would have been by nomadic hunter gatherers using hand axes and fire to clear small areas and then moving on. Several thousand years later Neolithic man began the systematic clearance of woodland by the slash-and-burn technique. In Wales one of the main areas of Neolithic settlement was in Pembrokeshire. An increasing population, vast flocks of grazing animals and improvements to tools as one passed to the Bronze, then the Iron Age and beyond, meant eventually only the most inaccessible of sites have not resounded to the axe, the crackle of flame and the smell of woodsmoke.

The Welsh laws of Hywel Dda provide us with information as to the forests of Wales between the 10th and 12th centuries. They contain one of the earliest references to tree planting in Wales, where every tree planted for shelter was valued at 24d. Eleven species are mentioned in the Laws, including alder, ash, crabapple, elm, hazel, oak, thorn and willow, with the value specified. Despite this interest and regulation the clearances continued unabated as the need for arable and pasture land increased, and the growing population required vastly increased quantities of fuel, and in 1386 a complaint was made over the felling of 3,000 green oaks and deterioration of the underwood through lack of care in Coedrath, which lay between Amroth and Saundersfoot.

In the Mabinogion we learn further of the forest clearances with the words: 'Dost see the great thicket yonder? I must have it uprooted from the earth and burnt on the face of the ground so that the cinders and ashes thereof be its manure; and that it be ploughed and sown so that it be ripe in the morning against the drying of the dew, in order that it may be made into meat and drink for thy wedding guests and my daughter's. And all that I must have done in one day.'

The antiquary John Leland, during his visit to Wales from 1536-1539, reported that western Carmarthenshire, and Pembrokeshire were largely denuded of wood, the latter being described as 'sumwhat barren of wood'. Of Strata Florida north of Tregaron, Leland wrote: 'Many hilles thereabout hath bene well woodid, as evidently by old rotes apperith, but now in them is almost no woode. The causses be these; first the wood cut doun was never copisid, and this hath beene a great cause of destruction of wood thorough Wales. Secondly after cutting doun of woodys the gottys [goats] hath so bytten the young spring [coppice regeneration] that it never grew but lyke shrubbes. Thirddely men for the nonys destroied the great woodis that thei shuld not haborow theves . . . Al the montaine ground bytwixt Alen [Elan] and Strateflure longgeth to Stratflure, and is almoste for wilde pastures and breding grounds, in so much that everi man there about puttith on bestes as many as they wylle paiying of mony . . . The pastures of the montaynes of Cairdiganshire be so great that the hunderith part of hit rottith on the ground, and maketh sogges and quikke more by long continuence for lak of eting of hit . . . a hille side Clothmoyne, wher hath bene great digging for leade, the melting whereof hath destroid the woodes that sumtime grew plentifulli theraboute'.

The poet Thomas Churchyard (1520-1604) in his best known work *The Worthiness of Wales* published in 1587 was moved to write: 'They have begun of late to lime their land And plough the ground where sturdy oaks did stand, They tear up trees and take the roots away . . . '

In 1980 Mrs Jean Buchanan and Martin Fuller produced a report on behalf of the Nature Conservancy Council (NCC) and WWTNC concerning ancient woodlands in Pembrokeshire. Although unable to cover all sites, including the Gwaun Valley, Tycanol, Penkelly and some eastern woodlands, they did bring together for the first time a wealth of information including reference to woodlands from the works of the early historians of Pembrokeshire.

The Marcher Lord of Cemais, George Owen (1552-1613) in his *Description of Pembrokeshire* provides us with much information concerning the state of the county's woodlands in the late 16th century, noting: 'This country groaneth with the general complaint of other countries of the decreasing of woods, for I find by matter of record that divers great cornfields were in times past great forests and woods '.

His list of lost sites is headed 'woodes and fforests in tymes past and nowe destroyed and arable lands'. The wealth of other detail on woodlands is remarkable, and gives us a precious insight into their state four hundred years ago.

Richard Fenton travelled widely in Pembrokeshire in the 1790s and, although his descriptions often lack precise detail, they nevertheless confirm the survival of some woodlands listed 200 years previously by George Owen, and the destruction of others. Of special interest is Narberth Forest, listed by Owen asn one of the 'best standing woodes' with red deer, 3,000 prime oaks, 11,000 fuel trees and 21,000 saplings. Less than a century later it had been reduced to a few small coppices and by 1790 Fenton saw it as '. . . great turn pike on Narberth Mountain, in my memory an open, dreary common, here and there shewing patches of stunted oaks, the grim remains of once a very flourishing forest; for even as later as the time of James the First it was by survey then taken, stocked with red deer, and contained eight hundred and seventy three acres of woodland . . . But the forest in that reign had begun to diminish and the wood to die away, cattle being suffered to browse its skirts everywhere, otherwise the ancient limits of the woodland much exceeded those assigned to it in the above survey, for the commons of Templeton and Molleston adjoining, and always esteemed part of it, had been entirely of the same forest quality, with that of the more preserved part referred to in the survey of James'.

By the Napoleonic wars there was a severe shortage of timber everywhere. Large landowners commenced tree planting, not only of native broadleaves; they also introduced species including conifers from elsewhere in Europe, Asia and North America. Although the practice continued during the 19th century, it was only on a small scale compared to our timber requirements, and much material was having to be imported. A crisis was reached during the Kaiser's war, when losses from enemy action curtailed imports and, as a result of this, together with increased demand, 182,900 hectares of woodland were clear felled in Great Britain. A direct result was the establishment of the Forestry Commission in 1919, with the task of setting up and maintaining its own forests and also of encouraging others to do the same. A further massive felling took place during the Second World War and, although much replanting has subsequently taken place, the accent has been on fast-growing conifers and only recently have greater efforts been made to promote the planting of broadleaves.

In Wales the loss of broadleaved woodlands during the present century has been critical, and more severe than in many parts of England or Scotland. In 1924, 24.5% of woodland in Wales was classed as 'felled or devastated' compared with 20.6% in Scotland and only 11.9% in England. The figure for scrub in Scotland was 19.4%, in Wales 13.8% and in England 5.4%. No less than 43.8% of woodland in Pembrokeshire and over 50% in Ceredigion had been felled. More recent investigations show that the trend has continued. Some 52% (1,330 hectares) of ancient woodland in Pembrokeshire has been converted to plantations or destroyed since the 1930s, and in 1985 only 1,244 hectares remained. The remaining woods are often under threat, not just from further felling or conversion to conifers, but often simply through a lack of management, usually an absence of open areas in glades, or rides, and a lack of fencing so that grazing prevents any natural regeneration. Unless urgent measures are taken, the wildlife value of more woodlands will further deteriorate during the first half of the next century. To both save and enhance Welsh woodlands, Cynefin, the Welsh contribution to the World Conservation Strategy launched in 1982, is prompting *Coed Cymru — Save Welsh Wildwoods* — a project sponsored by a group of organisations involved with farming, forestry and other activities in the countryside. 'The experience of the few must now be disseminated so that the majority of landowners and others interested in the future of our wildwoods can do their share and provide their contributions to the survival programme.' A five point plan has been initiated: 1. Coed Cymru — Save Welsh Wildwoods is a campaign to bring the task in front of the public. Exhibitions, television and radio items, public speeches, leaflets, practical demonstrations of woodland management and many other activities will inform and persuade the community to help save the wildwoods. 2. Existing advisory services will be co-ordinated to give expert help in each Welsh county and National Park and also on management of woodlands and in obtaining grant-aid. A telephone information service will be set up to help farmers and other woodland owners with technical and financial matters. 3. Discussions will be held with Members of Parliament and others concerned with policy and law making to examine means of facilitating the availability of financial aid and other assistance from government sources. 4. An educational service will be promoted through the schools to ensure that the wildwoods find a place in teaching programmes, especially in geography and biology lessons, and local studies. Emphasis will be placed on agricultural training, among Young Farmers' Clubs and in further education generally. 5. A number of general educational initiatives will also be promoted. Through Project Cynefin, initiated by the Prince of Wales' Committee's Environmental Education Group, schools throughout Wales will be collecting information about woodlands in their locality and youth groups will also contribute through the International Youth Year activities.

Education will ensure that the young people of today are better prepared to care for their woodland heritage when they reach adult life, and are in a position to take decisions about the future of our land.

A number of important woodland sites are already protected in Dyfed, including two in north Pembrokeshire. Penkelly Forest may well have been some 200 hectares in extent during Elizabethan times; now it is but a third of that size, though the old boundaries can still be traced. George Owen not only described the wood but also provided a valuable insight into its management when he wrote about 1600: 'The lorde hath a greate woodde there called the forest of Penkelly which lyethe betweene bothe the lordes demeynes called Henllys in the manor of Bayvill and Coort Hall in the manor of Egloserow, and the same contyneth of the usual measure of that contrey aboute 500 acres of wodde and is enclosed with quicksett and pale rownde about and under lock, and doth contyne in compasse nyne hundred perches, each perche being 24 foote in length, which maketh about 4 myles and thre quarters. And the lorde kepeth the same with all profittes thereof arising in his owne handes. Yt is all growne with great okes of 200 yeres growth and more and some young woodde of 60 yeres growth, and most of it well growen with underwoodde as orle, hazell, thornes, willowse and other sortes of underwooddes, the herbage whereof, yt beinge nowe enclosed, will somer 30 breedinge mares and winter 300 sheepe and 200 cattell well and sufficiently, beside swyne which may be kepte there. Allso there is in the said woodde 14 cockshottes wherein is greate store of woodcockes taken yearly, which cockshottes is the lordes owne to do therewith what he pleaseth. Allso there breedeth in the said woodde sparrhawkes which is the lordes also. Allso the panage of hogges, bees and hony, the herbage and all other profittes thereof is the lorde owne without any maner of parte or comon to be demanded by any tenannte or other. The herbage of the said woode is very good for cattell, horses, mares, sheepe and swyne and there is store of faire fresh ryvers and springes in the same woodde fitte and holsome for all kinde of cattell'.

He further said: 'Carieing woodde. Allso they are bound to cut, cleeve and leade home firewode for the lorde to burne at his manor howse out of the woodde or forest of Penkelly aforesaid. And the bayliffe for the tyme being of the said manor is bounde to admonishe and warne all the said tenanntes to do the said servies as well as abowt the hay, for the mille, and abowt the woodde as oftenn as neede shall require'.

There have been many changes since 1600, and unfortunately much of the structural diversity of the woodland has since been lost; management by WWTNC, which has a management agreement from the Forestry Commission, is aimed at recreating this. The nature reserve covers two broad habitats. Some 16.5 hectares is on well drained slopes dominated by sessile oaks and birch woodland typical of West Wales. This area has undoubtedly been managed over many centuries to supply bark for tanning, timber for pit props, buildings, fences and machinery, with small wood being converted to charcoal. Within the wood there are the remains of many charcoal hearths.

Much of the remainder of the reserve is underlain by glacial boulder clay deposits, the undulating terrain being poorly drained and dissected by many small streams. In these areas occur a variety of woodland types including plateau alder woodland and a variety of oak, ash, birch woodland containing both sessile and pedunculate oak, and a range of hybrids. Three trees of Midland hawthorn occur here, its only native site in West Wales, while other interesting species include bird cherry, aspen and guelder rose. The ground flora is also particularly rich, with wood millet, woodruff, woodland violet, water avens, adder's-tongue fern and large numbers of early purple orchids. Unfortunatey much of the woodland was devastated during the Kaiser's war, when a steam tramway was constructed to facilitate extraction of timber for use in the mines of South Wales. Few large trees now remain within the reserve area and the lichen flora and dead wood fauna is consequently impoverished.

During summer there is much to be seen. Pearl-bordered fritillaries, seriously declining on a national scale, can be seen in small numbers in the scrubby fields on the woodland edge, while later the silver washed fritillary is abundant. In one corner of the reserve some 20 wych elms support a small colony of the white-letter hairstreak, a butterfly totally dependent on elm, and which has suffered greatly in recent years from the loss of mature trees through Dutch elm disease. There are sunny glades where tall herbs and bramble provide a nectar source for these butterflies and a wide range of hoverflies, bees, wasps, beetles and flies throughout the summer.

The bird population has been restricted by the absence of nest holes and structural diversity in the young dense woodland, but in recent years has shown signs of improvement. Pied flycatchers have colonised the woodland during a recent westward expansion by this species, and in 1985 numbered about 10 pairs. During May and early June, redstart, wood warbler, blackcap, great, blue, coal and marsh tit can be heard in song, while the great spotted woodpecker drums on trees killed by competition with their neighbours. Sparrowhawks are frequently seen, while buzzards nest in the wood itself. Might we expect goshawks in the not too distant future in view of the colonisation of many counties by this fine raptor?

Foxes are common, using the wooded areas to lie up in daylight and also, no doubt, hunting the rabbits to be found in some of the drier sectors. Badgers use the wood for foraging; their paths can be seen radiating from four large setts on well drained slopes. Further evidence of their nocturnal activities are the extensive latrine pits along the woodland edge. Polecats are occasionally seen, and stoats and weasels are probably also present, preying on the numerous bank voles and wood mice. The rarest animal is the dormouse. This delightful animal has so for only been found in Pembrokeshire in the Gwaun and Nevern valleys, where a dense growth of hazel and brambles provides a plentfiul food supply before hibernation at the onset of winter. In Penkelly evidence of dormice has largely been gleaned, from finds of hazel nuts characteristically chiselled open to leave a regular hole with completely smooth edges. The streams and small waterecourses provide a home for frogs as well as a range of stoneflies, caddis flies and other insects. Brown trout and eels occur in the streams and migratory sea trout occasionally ascend to spawn on the headwaters within the wood.

The pied flycatcher takes readily to nest boxes and is extending its range in West Wales. (DF)

Although undoubtedly an ancient woodland, Penkelly forest has evolved to its present state after centuries of management by man, culminating in an almost total clear fell in the early 20th century. It is essential that the forest, in common with other Welsh woodland, be managed to ensure its healthy continuity into the next century and beyond. At Penkelly conservation and enhancement of the nature conservation aspects of the site are of overrriding importance, but commercial considerations need not be overlooked. The WWTNC has recently embarked upon an ambitious management programme aimed at maximising the natural history potential and generating some income, as well as developing an extensive footpath network to allow visitors to enjoy the many delights of the site.

Among tasks currently being undertaken is the creation of a system of wide rides and glades, so as to provide much woodland edge habitat essential to the ecology of many species, particularly butterflies and other invertebrates, dormice and birds. Within the main body of the wood, a programme of thinning has begun to promote development to a mixed age structure high forest system. In addition, management on a rotational system has commenced on one area of oak which in the past provided bark, charcoal and small timber. Some alder will occasionally be coppiced and a further area developed as a coppice with standard systems harvesting ash, birch, willow and alder underwood and promoting selected trees to produce timber.

About five kilometres to the south-west is Tycanol Wood, probably part of a large area known by George Owen as Killroth. The woodland is unique in West Wales for the way it overgrows numerous rocky tors. On the upper slopes are areas of birch-sessile oak and hazel, with a rather impoverished ground flora; lower down in damper areas is an oak-ash woodland with some alder. The site is noteworthy for its outstanding lichen flora, with no fewer than 330 different species, including many rare ones which festoon the trees, and carpet the boulders and rock outcrops. Many are old-forest indicator species. No fewer than 19 different epiphytic lichen communities and ten saxicolous lichen communities have been identified, containing about 20% of the total British lichen flora. Particularly exciting was the discovery of a species of lichen new to science in 1978 named *Cliostomum conranii*, the owner of the wood at that time being Dr Oliver Conran.

Some of the woodlands on the steep slopes around the Milford Havern estuaries, and in the Teifi Gorge on the border with Ceredigion, are of interest as fragments of ancient woodland. A notable species in these habitats is the wild service tree; of mainly south-east distribution it is virtually restricted in West Wales to these estuarine woodlands, where in autumn the rich leaf colour makes the few trees conspicuous.

The Cothi and Tywi valleys of Carmarthenshire support a number of fine woodlands dominated by strands of sessile oak. On more level terrain near to Llandeilo are two quite different sites. At Tregyb one of the largest blocks of woodland in south-east Dyfed is now owned by the Woodland Trust. Drier ridges alternate with flushes, and several stream valleys, supporting a mixture of ash and oak, alder and willow carr. Unfortunately much of the wood was clear-felled, so it consists mainly of younger growth; the only old trees which remain are along the boundary banks. Grazing here in the recent past has been minimal, so resulting in a well developed shrub layer.

Not far away is Castle Woods, beyond which lies Dinefwr Deer Park, one of the finest examples in Wales of a pasture woodland with oak trees of immense size and age. The lichen flora in the Deer Park and Castle Woods is of considerable importance and includes a number of extremely scarce species. To ensure their survival through a continuity of tree cover much effort has recently been devoted to tree planting. In the Deer Park this has been by trees planted singly in wire cages, so as to maintain the open nature of the habitat essential for light-loving lichens.

Fallow deer has been present at Dinefwr since at least Elizabethan times, but alas in the early 1970s the herd of Dinefwr white cattle was dispersed to another locality. There are only four other herds of ancient white cattle in Great Britain: those at Vaynol Park, Gwynedd; Woburn,

Bedfordshire; Chillingham, Northumbria and Cadzow in Lanarkshire. It is to be hoped that not many more years will elapse before management can be assured, so that they can be returned to Dinefwr, their rightful ancestral home.

Steep sessile oak woodlands are however much more typical of West Wales, and there can be few better places to view this habitat than in the Rheidol Valley. W.M. Condry tells us the place to enjoy the Rheidol is at Parson's Bridge. 'There is so much to see, so much to learn, by simply walking down that steep woodland path to the bottom of the gorge. There you meet with forest which may well be primeval: for not even the most rapacious have deemed it worth risking their necks to get at these crooked, stunted oaks and birches. . . . this dark, cool gorge with its ancient, mossy trees . . . and its deeply pot-holed riverbed far down between the rocks'. If you are lucky you may see a red kite soaring overhead, though a buzzard will be much more likely. A summer visit will certainly reveal birds like the pied flycatcher, redstart, and in the more open areas the tree pipit, easily recognised by its loud song and diagnostic song-flight.

White-letter Hairstreak

Steep sided sessile oak woodland typical of much of the Rheidol Valley and similar areas in West Wales. (NCC)

ABOVE: Pasture woodland, a scarce habitat in West Wales, at the Dinefwr Deer Park. (TMcO) BELOW LEFT: Ancient oaks in Dinefwr Deer Park, a habitat for rare lichens and rich invertebrate faunas. (TMcO) RIGHT: Tycanoe Wood, another ancient habitat. (SBE)

ABOVE: The Dinas reserve of the RSPB in the upper Tywi Valley. (TP)
BELOW: A section of the Dinas reserve with a ground flora dominated by bluebells. (TP)

ABOVE: Remains of an ancient birch forest at 550m in the upper Tywi Valley. (JPS) BELOW: Llanerch Alder Carr in the Gwaun Valley. (SBE)

ABOVE: Charcoal pit in Penkelly Forest. (SBE) BELOW: Estuarine woodlands on the Daucleddau. (SBE)

ABOVE: *Phaliota spectabilis,* a common but beautiful fungus often found growing at the bases of trees or stumps. (PC) BELOW: A number of sites in West Wales are renowned for their lichen floras, while most woodlands have a range of common species. (PC)

ABOVE: Laugharne fields; the ditches in areas like this are of high nature conservation interest. (JWD) BELOW: Well developed hedge and hedgerow trees. (NCC)

Farming Landscape of Dyfed

Gulls over ploughed field

Marion Shoard's *The Theft of the Countryside* might have us believe that agriculture has despoiled the rural scene. To the layman to agriculture and conservation this would seem a strange contention in Dyfed. Residents and holidaymakers alike cannot fail to be impressed by the range of landscapes, from the spectacular coast to the varied uplands, and the wildlife each habitat supports, but perhaps — in spite of this — is the balance right? Farmers, town and urban land users and policy makers have been much criticised, so you will have to decide where your sympathy lies.

Today Dyfed's 6,090 square kilometres of land includes some 4,900 square kilometres of enclosed agricultural land within which are some 750 square kilometres of rough grazings. Additionally, there are extensive areas of unenclosed common, much of which is upland — but not all. It is certain therefore that both wildlife and landscape in Dyfed are substantially dependent upon land which has also to provide a livelihood for those who farm it, and for those engaged in ancillary occupations. In Dyfed, farmers, their partners and workers, total some 19,000 people.

Farm Sizes in Dyfed

	1980	1982	1984
Under 10 hectares	2,380	2,749	2,594
10-20 "	2,087	2,077	2,018
20-30 "	1,746	1,661	1,595
30-50 "	2,420	2,340	2,299
50-200 "	2,186	2,247	2,317
Over 200	51	57	56
Rough Grazing only	191	208	389
Total Farms	10,967	11,339	11,268

Averages	1980	1982	1984
Farm size	41.9 ha.	42.0 ha.	40.7 ha.
Dairy herd	42 cows	47 cows	47 cows
Beef herd	15 cows	16 cows	41 cows
Ewe flock	180 ewes	182 ewes	186 ewes

The enclosed land extends to 300m AOD — here and there rather higher. Much land over 200m is affected by higher rainfall, steep slopes, thin soils and often stones and boulders, so that its agricultural potential is diminished. The climate predisposes a pastoral economy except on the coastal fringes, where most of the cereals and all of the early potatoes are grown. Of the 4,000 square kilometres declared as 'clean land', less than one tenth, 30,000 hectares or so, are in crops, which includes 21,800 hectares of cereals (mainly barley), 2,900 hectares in potatoes (2,300 hectares in earlies) and the remainder in forage crops for the county's livestock.

Dairy cows dominate the livestock economy with yields to the order of 4,750 litres per annum; this is below the United Kingdom average. The difficulty of producing good quality hay in the prevailing climate is a major factor here — but increased and improved silage production could change this. Some 39,000 beef cows are kept — mostly in upland areas; the calves may be moved on for fattening. Breeding ewes total 819,000 and 90% of them are in the original and extended Less Favoured Areas (LFAs) — most in Carmarthenshire and Ceredigion. The influence of the hill ewe is seen in a lamb crop of 94%. Pigs, poultry and horticulture play less important parts in the overall farm economy.

Dyfed Agriculture Statistics 1975-84

The following is an abridged summary of the statistics relating to stocking and cropping (areas in hectares).

	1975	1980	1982	1984
Cereals	30.405	27,100	23,012	21,787
Potatoes (total)	2,836	3,000	2,915	2,946
Potatoes (early)	2,235	2,300	2,291	2,310
Fodder Crops	3,950	3,470	2,768	2,393
Other crops and Fallow	1,367	840	1,591	795
Temporary Grass	57,371	60,140	62,040	59,719
Permanent Grass	266,809	276,600	278,400	284,977
Total Crops and Grasses	362,738	371,140	370,726	372,948
Rough Grazings (excluding Commons)	87,776	71,570	69,535	70,213
Overall Total including Woodlands	462,230	458,740	457,188	459,383
Regular Workers Wholetime M & F	4,392	2,934	2,897	2,486
Total Workers	8,840	7,659	7,630	6,838
Total farmers and workers	18,841	19,533	19,684	19,483
Dairy cows and Heifers in milk and Cows in calf	177,550	186,953	197,721	209,409
Beef Cows and Heifers in milk and Cows in calf	51,704	43,314	41,183	39,477
Heifers in-calf 1st calf - Dairy	27,648	30,113	32,041	32,589
Heifers in-calf 1st calf - Beef	7,094	4,861	4,833	4,658
Other cattle and Bulls	280,799	264,554	266,069	276,505
Total cattle and Calves	547,860	529,795	541,847	562,638
Sows and Gilts Breeding	6,064	4,325	4,994	3,974
Total Pigs	36,978	30,375	31,342	29,535
Total Breeding Ewes	627,187	743,993	750,718	819,116
Total Sheep and Lambs	1,165,187	1,445,363	1,516,116	1,678,550
Total Poultry	2,958,926	2,276,886	2,106,330	483,298

Trends Summary: *Reducing* - cereals, fodder crops, number of farm workers, beef cows, beef heifers, sows and gilts, poultry and rough grazings. *Static* - Early and main crop potatoes, temporary grass, total farmers, partners and workers and 'other cattle'. *Increasing* - permanent grass, dairy cows, dairy heifers and breeding ewes.

In 1984, 11,268 farms of various sizes and types in Dyfed made returns to the Agricultural Census; the table compares the statistics with those for 1980 and 1982. The larger and medium sized farms have generally re-equipped with new buildings for their chosen farm system. The small and part-time farm has often not been able to afford radical improvements. Thus most larger dairy units are equipped with parlours, cubicle housing and self feed/easy feed or mechanised handling silage systems. Silage and cubicles are also now found on the more substantial beef farms. Grain silos, potato sprouting sheds and irrigation reservoirs are common on arable farms. More recently, sheep in-wintering and lambing sheds have been built on upland holdings. Naturally with limited labour, and that often family labour, most farms have adequate modern machinery, but there are a growing number of contractors. There is however, still considerable co-operation between neighbours in the ownership and/or use of machinery — especially with gang work. Dyfed farmers have always been ready to adopt new technology, high nitrogen for grazing dairy cows, wilting and additives for silage, computers in the milking parlours and farm office, automatic slurry scrapers, irrigation and the plastic covering of early potatoes, intensive silage beef, pregnancy scanners for breeding ewes and many other developments.

The bare statistics describe the working parts of the county's agriculture industry, which of course has the prime objectives of producing food and providing a livelihood for the rural population. Happily, the result is a beautiful and varied landscape which, as well as significantly contributing to the national larder, still provides habitats for much of our wildlife heritage.

There are many visual contrasts in the landscape. In parts of the south western coastal area, intensive early potato production with its tidy closely managed husbandry is seen among rough and wild cliff-top areas — interspersed with wooded valleys and pastoral lands. Enclaves of arable land occur often with larger enclosures than adjoining pastures; glance at the coastal fields north of Aberaeron and contrast them with the nearby upland. Sea cliffs and adjacent pastures are vital to the wellbeing of the chough. Dyfed is a stronghold for this red-legged crow, and it seems likely that where sheep and cattle are grazed they have the best habitat for survival.

The hinterland pastures and small woodlands gradually give way to the hillside *ffridd* just before the open hill. In the *ffridd*, a mosaic of bracken, bilberry and mixed old grassland often rich in plants has survived, because of slopes and difficult access. Many valuable habitats have survived where banks are steep or the ground is boulder-strewn or badly drained; here improvements would be impracticable, uneconomic or just impossible.

The strong and fertile pastures of our lowlands are perhaps ordinary in wildlife terms. They are, however, the backbone of our dairy farming enterprise - with some beef cattle and sheep. Here sound hedges with well-grown trees, mainly ash, oak, sycamore and some beech, are found.

Old farm trackways, farm quarries, some with the benefit of open water, old walls and buildings, are all part of the scene and are often valuable in wildlife terms. On many old farm buildings ferns, like maidenhair, spleenwort, wall-rue and even rusty-back, enjoy a secure root-hold in the often lime-rich mortar. Old buildings often provide nest sites for barn owls and this valuable farm pest control agent could be better encouraged by nest boxes, which are now in fact increasingly provided for them in modern farm buildings. The hedgerows and stone-earth banks, often with associated ditch and fence line, have for centuries provided link-ways for wildlife as well as field boundaries and stock-proof barriers.

Rough shooting for rabbit, pigeon and wild game is popular, with some farms and estates run as game shoots, mostly for pheasants and wildfowl. Management of hedge, field and woodland for the shoot is often of great benefit to wildlife. Estates with parkland are also of impressive scenic value.

Marsh ditches in flat lowland areas like Laugharne, Pendine, Pembrey and Castlemartin are rich in plants and aquatic invertebrates - dragon and damselflies for instance live amongst marestail, frogbit and other plants of interest. Many farm ponds have been lost over recent decades, but some survive, and at present farmers are interested in creating new features of this type. Often they have a multiple role in that wildlife, amenity, fire fighting, fishing and shooting can be part of their *raison d'etre*.

Irrigation reservoirs have been constructed on nearly all early potato farms; this is a major contribution to wildlife, although it must be admitted that abstraction in May and June limits their overall value; nevertheless migrating waders, wintering wildfowl and other aquatic creatures have benefitted. The original engineering concepts of regular tank-like shapes are giving way to varying depths and featured shorelines with spits and bays, islands are frequent, and rafts have been placed in some, so providing safe nesting and roosting sites.

Hedgerow removal has not been a serious environmental issue in Dyfed. Where many small fields existed, some rationalisation has been essential to farming, with no great loss to conservation. Perhaps hedge management has not been of a high standard everywhere. There are many kilometres of hedgerow neglected and eroded and no longer stockproof, and such hedges have lost a great deal of their value for wildlife. On the other hand these untrimmed hedge remnants offer mature hawthorns which bear berries for winter birds and, at their base, secure homes for small mammals. Hedge laying seems to be reviving as a technique as does repair and renovation of selected stone-earth banks. There are farms where field re-planning has led to the removal of some old enclosure boundaries and the planting of new hedge-lines. Hopefully ancient and species-rich hedgerows will be retained; these are often found as boundaries to farms and along lane sides. Where hedgerow trees have been lost, not least by the ravages of Dutch elm disease, tagged saplings are being selected for survival and new young standards or whips are frequently seen. Peculiarly, there occur in north Pembrokeshire and south Cardiganshire hedges of laburnum, which make a superb spectacle when in full flower. Could it be that this species resisted early rabbit damage, and thus the hedges flourished where other species might have failed?

Farm woodlands may well take on a new lease of life with the timber managed for firewood production by a revival of coppicing. The new Forestry Commission Broad-Leaves Grant Scheme is already being taken up by many. Fencing out stock to ensure regeneration, planting of clearings within the woodland and other improvements can only help. The recent limitation on unlicensed timber cutting on the farm is welcomed by conservationists. Corner plantings are also being carried out, though not everybody likes these, and some say that an excess of corner plantings could lead to a 'lollipop' landscape! Steep banks and the less accessible corner are often covered by scrub, with gorse and blackthorn the dominant species. Such areas are the haunt of stonechat and linnet, while blackthorn is the food-plant of the brown hairstreak. Fewer than 150 colonies of this attractive butterfly now exist in Great Britain, with West Wales one of its strongholds. In wet areas willow and in places alder carr support rich insect faunas.

Arterial schemes on rivers and lesser channels, field drainage and reclamation and the improvement of heather moorland to grasslands have all contributed to habitat loss. Yet much of the field drainage carried out since the last war involved replacement rather than pioneer work, of old stone and clayware systems laid often in the 'golden years' of agriculture in the last century. Intensive livestock farming, particularly dairying, has led to pollution of ditches, watercourses and rivers, as well as of ponds. Cow slurry, 'brown water', and silage effluent, which is particularly damaging because of high biochemical oxygen demand, have all taken their toll of water life, and at times serious loss has occurred to valuable fisheries, quite apart from problems caused to water treatment works. The careless disposal of sheep dip has also caused some pollution incidents.

Pastures unploughed for many years still occur and are of great value to wildlife. Part of the sheep grazed ranges at Castlemartin abound with green-winged orchids and field gentian in due season. The open field strip systems at Laugharne are cut late for hay and then grazed. In south-east Carmarthen are the *Carum* meadows, while in central Ceredigion the rhos grasslands are an unique habitat. Here several wet sedge-rich pastures are nature reserves, where the traditional late summer

ABOVE: Fign Blaen Breifi nature reserve. (DRS) CENTRE LEFT: Western gorse and bell heather on Dinas Mountain. (SBE) RIGHT: Oblong-leaved sundew at Cors Caron. (JPS) BELOW LEFT: Marsh clubmoss at Brynberian. (JWD) RIGHT: *Sphagnum recurvum* at Cors Caron. (JPS)

PLATE I

ABOVE: Castle Woods and Dinefwr Castle. (SHR) CENTRE LEFT: Bastard balm at Coed Devonald. (JB) RIGHT: Midland hawthorn at Penkelly Forest. (JC) BELOW LEFT: Wood anemone. (JWD) RIGHT: Spindle at Sam's Wood. (SBE)

PLATE II

to early spring grazing is still maintained. Blue with devils-bit scabious in the late summer, these pastures are a stronghold for the marsh fritillary butterfly, which has disappeared from many other sites in southern Britain due to changing agricultural practices.

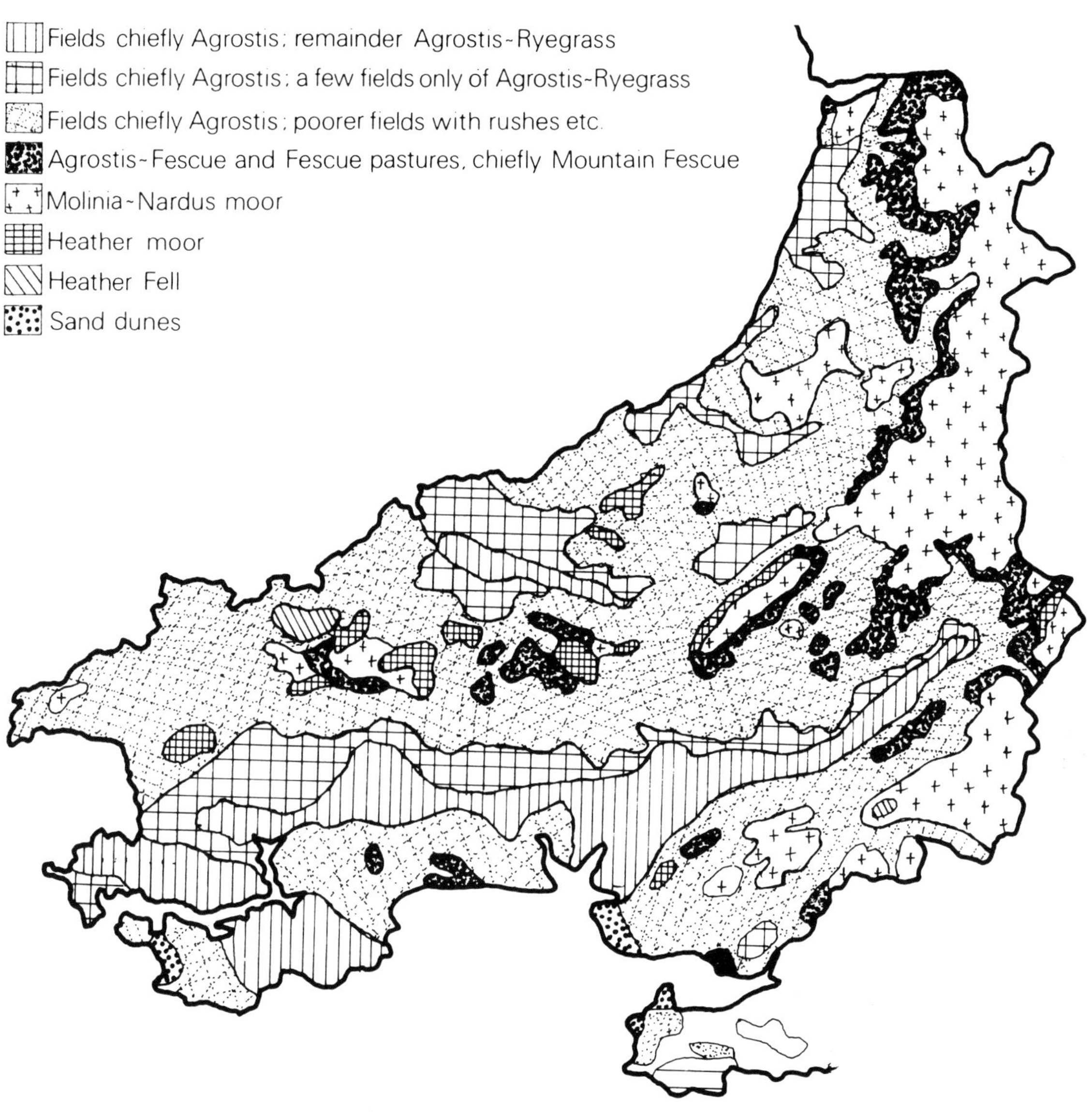

Grasslands of West Wales, (adapted from Stapledon, G. & Davies, *Grassland Survey of England and Wales*, Faber 1940).

Intensive grassland systems with heavy grazing pressures and several cuts of silage each season must have contributed to the decline of corncrakes and grey partridge. The corncrake last bred in Pembroke in the early 1960s; now it is but a scarce passage migrant. Paradoxically however, it is probably the underuse of common land grazings that has caused scrub encroachment and dense herbage, affecting the value of such areas for nesting curlews and lapwings. Uncontrolled burning of these rough grazings can be a serious hazard to wildlife, but carried out within legal periods and safely it may benefit the habitat in, for example, *Molinia* wetlands and heather moors.

Some upland farm tracks have in recent years scarred the Dyfed landscape. The follow-up plateau improvements are beneficial for cattle and sheep, and could be in conflict with wildlife and landscape. Will upland birds and plants be ousted by improved upland grazings? To some, even bracken is pleasing, particularly because of its splendid autumn colours; it is only since cattle grazing has given way to sheep that bracken has become more dominant. It is, however, often the only vertical component in the uplands with an edge feature, and it is favoured therefore by insects, birds like the whinchat and small mammals. Upland drainage, hill gripping, is not a frequent practice in Dyfed, so erosion and consequent silting of streams and small open waters has not proved a problem. Acid rain in the uplands may well be a problem in parts of Dyfed but it cannot be 'laid at the door' of modern agriculture!

Farm wildlife habitats need to be managed sympathetically for their conservation value, while at the same time the farmer earns his living, for example, not trimming all the hedges at one time, trimming them late to leave berries for birds, not cleaning a pond in one operation, not to use sprays up to and into the hedge base, and perhaps not planting trees in areas already of great wildlife value. Hopefully these practices are being considered and taken up by many farmers in Dyfed.

The recent innovation of the plastic covering of early potato and carrot crops results in short-term lost amenity, as shimmering artificial lakes. Plastic in hedgerows and the black noxious smoke from burning used polythene is also offensive. There is a need to cart spent plastic mulch to official tips for disposal.

Recent changes in the Welsh Office Agriculture Department (WOAD) conditions and levels of grant aid in the Agriculture Improvement Scheme (AIS) will reduce land reclamation, and assist in providing better storage and disposal methods for farm waste materials. Hedge planting, shelter belts, traditional walling, and tree preservation and planting are being encouraged. Indiscriminate grass and heather burning is controlled to prevent damage to wildlife and especially nesting birds at a critical time of year.

Although heavy dressings of nitrogen fertilisers are applied on dairy and potato farms, nitrogen pollution rarely occurs. Nitrogen does not leach significantly from grassland, and the mainly fast-flowing rivers do not allow nitrogen to accumulate in damaging concentations. Hedge removal and cereal stubble burning is not a serious problem in Dyfed. Seed dressing and pesticides are not regarded as a threat to habitat in most situations, because of the predominantly pastoral nature of the farming.

The present WOAD regulations and grant-aided schemes are environmentally positive, and provide a carefully balanced approach to reconciling the interests of agriculture and the environment. Protection to special habitats is provided by Sites of Special Scientific Interest (SSSIs) which are increasing in number and carefully administered by the Nature Conservancy Council (NCC). The Pembrokeshire Coast and Brecon Beacons National Parks control planning approvals on many farm improvement items, and have to sanction all grant-aided schemes carried out within their boundaries. Areas of wildlife or scenic value are being protected by management planning agreements where farmers are financially assisted to manage their land sympathetically. The Welsh Water Authority (WWA) also has to be notified of all sensitive developments affecting water courses before work starts, as a condition of grant payment. Disposal of silage effluent, livestock waste and sheep dip is covered by this provision.

Dyfed, through the Dinas Conference of 1972, in the Upper Twyi Valley, has important early links with Farming Forestry and Wildlife Advisory Groups (FFWAGs), for it was from this Conference that forestry joined with farming in the conservation movement. Derek Barber (now Sir Derek) spoke with alliterative skill of sheep farming, spruce plantations and sparrowhawks. FFWAGs have been set up on a vice county basis in Dyfed, as an Agricultural Development and Advisory Service initiative, to foster a better understanding of nature conservation and its interaction with commercial agriculture and forestry. Farmers, growers, foresters and many of the related official agencies have been represented at discussions, meetings, shows, farm walks and demonstrations. A comprehensive advisory service is offered by members and paid advisers may in the future be employed to further the work.

Agriculture is experiencing the challenge of a period of low profitability as a result of its tendency to over-produce for limited markets. The Common Agricultural Policy is being modified to tackle this problem and quota impositions and price restraint currently, and in the future, are likely to have wide-ranging effects. New ventures such as deer farming, goats and sheep for milk, farmhouse cheese production, farm ice cream, yoghurt, milk chocolate and others are being adopted, but it is unlikely that the basic enterprises will be substituted on any scale for a long time to come. Rural diversification into tourism, craft and light industry is being encouraged and indeed, some of the new 'high tech' enterprises may be based or run in the more rural districts. Farm tourism is becoming increasingly important to the rural economy, especially as traditional farming incomes come under pressure. As well as being a noble, some would say essential, objective in itself, preservation of the environment has an economic benefit. Tourists can be offered a more interesting and fulfilling holiday in a beautiful and wild landscape which is sensibly integrated with agriculture. For local inhabitants, it is altogether more satisfying to live and work in amenable surroundings.

Farmers and growers have shaped and developed our countryside and are its main custodians. It is they, with their labour, livestock and machinery, who have created our rural environment — and in practical terms only they can sustain it. We must also never forget that we need to eat, and agriculture's prime role is to supply that need.

Pheasant and Corn Marigold

ABOVE: Rhos Pil-bach nature reserve; visitors examining the flora of this unimproved wet pasture. (JPS) BELOW: Large irrigation reservoirs in Pembrokeshire are particularly important for winter wildfowl; this one is at St Ishmaels. (JWD)

ABOVE: Cowslips have disappeared from many pastures but still flourish on some clifftops. (NCC) BELOW LEFT: Adder's-tongue fern. (NCC) RIGHT: Lapwing. (HP). Both species have vanished from many areas as wet pastures have been improved.

ABOVE: Cors Fochno looking north-east with the Leri in the foreground. (NCC) BELOW: Cors Caron looking west; traces of the peat diggings may be seen in the foreground and the course of the Teifi in the centre. (NCC)

Mire, Fen and Heath

Snipe Curlew

Most who travel west from Carmarthen by train hardly notice, within a few minutes of commencing their journey, that there is a heathy area known as Cors Goch extending on either side of the track. Here the line passes over history for, after the Tywi changed its course, a shallow lake remained to the west of what is now Llanllwch village and the United Counties Show Ground. Gradually this infilled as a succession of aquatic plants dominated the open water, until these in turn were replaced by bog mosses, *Sphagnum* species. Over the past 10,000 years these humble plants, the main constituent of acid peat, have lived and died here, gradually building up the peat layer so that it is now up to five metres deep. The construction of the railway line in 1854 was soon overcome by the Victorian engineers, who provided a foundation of brushwood and bales of wool on which the ballast and tracks were laid. Spare a thought for this when you travel across the mire by train.

Surveys show that 84% of the total area of raised mires in Great Britain have been lost or seriously degraded through afforestation, agricultural reclamation, peat winning or burning since 1850. Cors Goch is one of the most south-westerly raised mires in Great Britain, and one of only six large examples of this type of habitat remaining in Wales. The general shape of a raised mire is that of an inverted saucer. The high centre is dominated by the bog mosses; an actively growing terrain of deep pools in all shapes and sizes, *Sphagnum* filled depressions and domed heathy hummocks. Move towards the edge where, in the slightly drier conditions, purple moor grass becomes the main species with at Cors Goch some fine stands of bog myrtle, the aromatic leaves of which make any visit there so memorable. At the very edge, where the mire meets the valley sides, are other wetland plants like marsh marigold, yellow iris and reed canary-grass.

Peat mires are storehouses of pollen grains deposited here over centuries, for the outer shells resist decay and survive unchanged in the oxygenless peat. The different species each seem to have their own considerable beauty and unique shape when viewed through a microscope, and may be recovered from sample cores taken from the peat layer. From these, the vegetative history of this part of Carmarthenshire has been revealed. The initial disappearance of trees, the first cereal pollen dating from Neolithic farming, the changing tree patterns between the 18th and 19th centuries, even the onset of coniferous plantings, can all be found in the pollen record.

A section of Cors Goch has alas been lost to rubbish tipping, and nature conservationists had a struggle lasting several years until late in 1981, when Carmarthen District Council finally abandoned its plans to extend the tip over much of the northern side. Areas at both the east and western ends south of the railway line are owned by the WWTNC. The western end has been the scene of drainage efforts in the past, and a series of ditches cross the site. As a result this has lost much of its hummocky surface and pools in the central area, and in slightly drier conditions birch, alder and willow invasion is a serious problem. Scrub removal by hand, and the blocking of the ditches so as to raise the water table, are conservation measures currently in hand.

A much larger raised mire is Cors Caron just north of Tregaron, a National Nature Reserve (NNR) managed by the NCC who own a large part of the site. This was the first true raised mire to be described in detail in Great Britain, and so is a classical site where the development sequence is well documented. Educational facilities are provided for, with a trail across the south-east part of the reserve. At one time Cors Caron was an important source of peat for fuel, and extraction only ceased about 200 years ago. The West Bog beyond the Afon Teifi escaped much of this peat digging and is the best preserved part of the Reserve. Much research is carried on at Cors Caron into the vegetation, hydrology and the nutrient status of the peat, and into population studies of several insect groups, birds and otters.

The large heath, one of our most variable butterflies, reaches the southern extremity of its range in Great Britain at Cors Caron. It can be seen on the wing from mid-June to late July, and overwinters in the larval stage, the white-beak sedge its main food plant. About 70 species of bird breed on the reserve or in its immediate surroundings, while rare visitors have included purple heron, garganey, Montagu's harrier, quail, corncrake and great grey shrike.

Another classic raised mire is Cors Fochno or Borth Bog, which occupies an estuarine site on the south side of the Dyfi estuary. Alterations, mainly for agriculture, have reduced it to about a third of its original size; nevertheless it contains one of the most extensive tracts of unmodified raised mire in Great Britain. The rosy marsh moth, discovered at Yaxley in the fens in 1828 and subsequently at other East Anglian sites, became extinct there by 1851 with the draining of Whittlesea Mere. One can imagine the excitement in 1965 when a large colony was discovered on Cors Fochno, where its food plants, the bog myrtle and bog rosemary, are abundant. Another interesting resident is the black grouse, the small flock being almost unique at this low altitude in southern Britain. A small number of usually less than 100 Greenland white-fronted geese winter here, feeding on the mire and roosting on the nearby estuary. Although preferring raised mires, in Great Britain this goose in recent years has taken more to feeding on grassland and arable land, a habit long adopted by its Siberian race.

West Wales has good examples of other mires. There are several basin mires which have developed in depressions where the water table is stagnant. At Beacon Bog, a site of some 11.5 hectares near Carmarthen, bog rosemary grows at one of its most southerly locations in Great Britain. Other typical mire plants here are the royal fern, cranberry and petty whin. In central Ceredigion the end of the Ice Age between 13,500 and 10,000 years ago left many examples of relict pingos, formed by ice-covered mounds with the permafrost. As the thaw proceeded, so debris from the mound formed low ramparts, trapping pools of meltwater. There are excellent examples in the Cleddyn valley not far from Lampeter, and two in the WWTNC Rhos Glyn-yr-Helyg reserve close to the Afon Grannell in the same area. Both of these contain actively growing basin mires with a rich flora, and are now fenced so as to prevent stock becoming trapped in the soft surface.

There are also a number of flood plain and valley mires in West Wales, that on the upper reaches of the Western Cleddau near Mathry being one of the largest left in Wales. Part is a NNR, owned by the NNC, while slightly upstream of this is a smaller section owned by the WWTNC. Charles Hasall, writing in 1794, mentions that this together with several others 'may be easily drained and would make the finest meadows imaginable, the owners patiently receive the small rents they now produce without seeming to entertain an idea of increasing their rentals by draining their fens'. Fortunately for us, the core of what Hassall viewed near Mathry, remains reasonably intact. A feature of these valley mires, which Pembrokeshire shares with Anglesey, are the stands of tussock sedge, many of prodigious proportions, standing up to two metres high and three metres across in ungrazed situations. They provide a superb lying-up area for otters, and perhaps even the occasional breeding site for this declining mammal. At the Western Cleddau there are also scarce plants; the bay willow occurs at its most southerly location in Great Britain, and marsh fern, royal fern, great fen sedge, greatwater dock and wavy St John's-wort are others to cause excitement for the naturalist.

LEFT: Round-leaved sundew. RIGHT: Wavy St John's-wort.

Bog rosemary.

In Great Britain, among the most threatened habitats in recent years have been the lowland mires, fens and marshes. A major contributory factor has been improvement in machinery, together with financial incentives to convert more land to agricultural use. Once the water table has been lowered, there is little chance of reversion to the original wetland state. Between 1910, when there were 711 fens, valley, basin and raised mires in Wales, and 1978, 422 had been lost, and many of those remaining had been interfered with in some way, and were still under threat.

At one time fens occupied most of our river valleys; now only fragments remain here and around the margin of the few lowland lakes in West Wales. George Owen has left us with a tantalising insight into some of the birds of the Pembrokeshire wetlands in the 15th century when he says: 'In the bogges breedeth the crane and byttur, the wild duck, the teale, and diverfe others of that Kynde; on high trees, the heronfhewes, the . . . fhoveler & the woodquiftes'.

His reference to the 'fhoveler' nesting in trees must surely be to the spoonbill. He goes on to tell us that herons can be enticed to nest in trees by 'placing horefehead bones' upon the branches. Perhaps this is something that might be tried at selected spots by 20th century conservationists?

Ivy-leaved bellflower.

Reed warblers have colonised many West Wales reedbeds during the last 15 years. (DF)

Two centuries later Hassall was writing that 'the fenny parts of Pembrokeshire are not extensive.' and tells us how Castlemartin Corse, 'a tract of several hundred acres; which, within these four years, was a perfect bog, of little or no value. Mr Campbell of Stackpole Court having obtained an act of parliament for draining and inclosing it, such an improvement is already made as must convince every observer, what ground of this sort is capable of, under skilful and attentive management.

A 20 hectare fen owned by the National Trust is now all that is left of the wetland here and includes six different fen and fen-meadow plant communities. Several scarce species are recorded, including fen pondweed, slender sedge, long-stalked yellow sedge, marsh helleborine, lesser water parsnip and black bog rush. The great green bush cricket is abundant here. It has a coastal distribution in West Wales from the Marloes Peninsula to around Tenby, and its strident song can be heard from the early afternoon until well after dark, from mid-July to the first frosts of October.

On what is now Goodwick Moor, a memorable battle took place in 1078, when the forces of Trahaearn ap Caradoc of North Wales routed those of South Wales commanded by Rhys ap Owen. One may assume from this event that the area was more open than today, and indeed in 1794 it was described by Fenton as an 'extensive moory flat'. The construction of a road along the edge of the beach, and a railway embankment during the development of Fishguard Harbour, impeded drainage so that reed beds and fens developed. Although the embankment has been removed, and a large part of the moor used as a rubbish tip and subsequently sports fields and garages, the eastern section remains, owned by Preseli District Council and leased as a nature reserve to the WWTNC.

The nature reserve contains an exciting range of wetland habitats, reedbeds which at the northern extremity are still influenced by salt water at high tides. There are fen, marsh, carr and some small areas of open water in ditches and recently hand-dug pools. Other recent management has included the installation of a sluice, with the objective of maintaining a high spring and early summer water table, so as to promote a healthy reed bed and suppress invasion by willow and bramble: A boardwalk has been constructed so as to allow visitors to view the very heart of the site. This includes a small area of bog myrtle, which Fenton noted as 'covering the whole vale'. Now it is confined to the southern boundary, where invading willow has been removed by successive work parties of WWTNC volunteers. There can be few more glorious wetland plants than bog myrtle when its golden catkins are in full bloom in April and early May.

Of other wetland sites, there are estuarine reedbeds like those at Slebech on the Eastern Cleddau and on the Teifi just upstream from Cardigan. Near Tenby, Penally Marsh has been visited by botanists since the last century, though alas both great fen-sedge and greater spearwort have now disappeared. Over the border in Carmarthenshire, the Witchett Pool in the MOD range near Laugharne has extensive fringes of reedbeds. At Falcondale Lake near Lampeter there is a wide zone of reed sweet-grass at its only known site in Ceredigion. Talley Lakes formed as the result of glacial action exhibit a well marked sequence from open water, through reed swamp, to mature alder carr.

Areas of lowland heath in West Wales are mostly confined to the St David's Peninsula, the coastal region west of Goodwick, the lower reaches of Mynydd Preseli, and a few fragments on the Ceredigion coastal plateau. Many will have formed as the result of early forest clearances, followed by soil deterioration due to leaching. Several early historians, although referring to the rich arable lands around St David's, make no mention of the heaths which today are such a significant feature of the landscape. Perhaps not exciting to the casual onlooker, they are nevertheless of great interest to the naturalist, and contain a diversity of habitats from dry heaths to valley mires and open water.

Barely two kilometres north-east of St David's is Dowrog Common, now owned by the National Trust, but managed for many years as a nature reserve by the WWTNC. To the north Dowrog merges with Tretio, while to the south is Waun Fawr, which brushes the outskirts of St David's itself. Being so close to the City seems to have attracted the attention of naturalists for well over half a century, and consequently we know more about Dowrog than possibly any other.

It extends for about 83 hectares, with wet and dry heath, grass heath, tussock sedge fen, willow carr, marsh, open water and a section of the River Alan. Formerly grazing would have been important, as those with common rights turned their cattle and horses out to feed. As this practice ceased, so heather and gorse became dominant and the development of scrub was only curtailed by casual burning. The open water on Dowrog also results from human activities. Several small pools are where clay was extracted for caulm, while another is marked on the map as a gravel pit. Of special interest is Dowrog Pool, formerly of some seven hectares, but with a depth rarely exceeding a metre. Its most likely purpose seems to have been a holding pond for water brought by a leat from the River Alan, and so to the mill at Rhodiad slightly downstream. Alas, during the past quarter century, it has been sadly infilled, mainly by bulrush, so that rather little open water exists here in summer.

Nearly 300 plant species have been recorded on Dowrog Common, including a number of particularly scarce species. Fibrous tussock-sedge fringes the western margin of Dowrog Pool, at one time its only known site in Wales, until recently when two further colonies were found around heathland pools near to St David's. At the edge of shallow pools, pillwort and the delicate white flowers of floating water-plantain may be found. The paths that cross the heath have a special significance for two scarce plants. As you splash along in the early months of the year, the bright green leaves of three-lobed crowfoot attract attention; at the other end of the season when these same tracks are dry in late summer look for yellow centaury. Dragon and damselflies abound in the summer, while in winter small numbers of waterfowl, including occasional parties of whooper and Bewick's swans, visit the pools. Hen harrier, merlin and short-eared owls are winter raptors, while in days gone by the Montagu's harrier nested here. After many years when no young were reared in Great Britain, this summer visitor now seems to be faring better, so that it may return to old haunts. Will these include Pembrokeshire?

The management of lowland heaths like Dowrog Common is the key to their continual well being. Most were much more open in the past, probably primarily grass heath and, as a result of the changes, some plants have already been lost. Much has been learnt over the past ten years from the management initiated at Dowrog, and hopefully this can be translated to other nearby sites. Tracks have to be kept open, the ideal method by pony trekking on well defined routes. Grazing needs to be re-instated, and at Dowrog the interest and co-operation of an adjoining farmer, I. Jameison, who is also an honorary warden, has made this possible and much invaluable experience gained. Cattle grids will, however, need to be installed to enable this to be continued beyond the current experimental stage. More open water has been provided by the excavation of a series of pits and pools, the cost of which was met by a Welsh Water Authority grant. Volunteers have also been busy at Dowrog Pool, hand pulling bulrush in an effort to retain some open water among the rapidly encroaching vegetation. Burning of heathland has in the past been haphazard, and so firebreaks have been created to reduce the danger of deliberate fires, while in addition small areas are now burnt on a long term rotation. The future for these lowland heaths is, as a result of these activities, now much brighter than at any time in the past half century.

Common Toad with Marsh Marigolds

ABOVE: Goodwick Moor; the nature reserve is on the east, bounded by the Drim brook and the main drainage channel. The outlines of old ditches are clearly visible. (SJ) BELOW: Brynberian Moor. (SBE)

Habitat management on the Dowrog Common reserve. ABOVE: Horse grazing. (SBE) BELOW: Burning firebreaks. (JWD)

ABOVE: *Sphagnum* pools and hummocks. (JPS) BELOW LEFT: Bogbean. (NCC) RIGHT: Royal fern.

ABOVE: Reed beds and board walk Goodwick Moor. CENTRE LEFT: Fibrous tussock-sedge at St Davids. RIGHT: Marsh fern at Penally. BELOW LEFT: Pillwort at Dowrog Common. (All SBE) RIGHT: Wavy St John's-wort at Dowrog Common. (JWD)

PLATE III

ABOVE: Whorled caraway meadow near Llanddarog. LEFT: Marsh pea at the Ffrwd Farm Mire nature reserve. RIGHT: Fen orchid, formerly found at Tywyn and Pembrey Dunes. (All RP)

PLATE IV

ABOVE LEFT: Lesser butterfly orchid. (JH) RIGHT: Greater butterfly orchid. (JH) BELOW: Marsh marigold. (NCC)

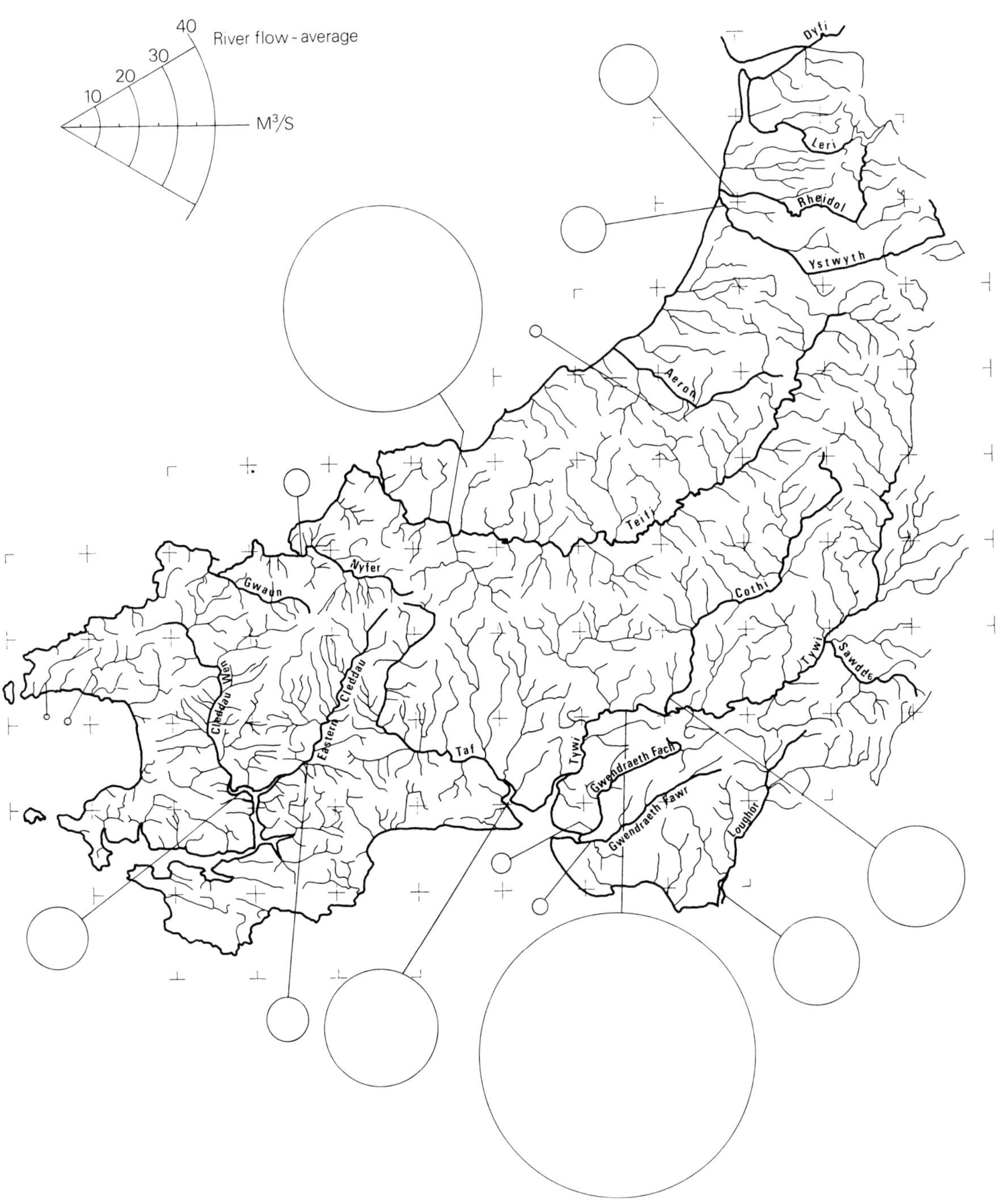

Rivers and major tributaries of West Wales.

Open water, rivers and streams

Cenarth Falls, River Teifi

The open water of lakes, ponds, rivers and streams represents the most widespread of the natural and semi-natural habitats still remaining in West Wales. Most of our fresh water resources show the mark of man's attention; only the deepest parts of our lakes escaped his activities.

Against this intereference can be set the dynamic nature of the water world. Rivers are not constant environments. The volume of water sediments carried by the river is forever changing. The 'energy' represented by this load varies, forcing the river to respond by changing course and hence causing the removal of deposition of banks and river floor. Lakes and ponds are constantly being filled with sediment by incoming streams. The development of aquatic vegetation reaches its climax with the disappearance of the water body under willow carr or alder which, if unchecked, will revert to woodland, or in particularly wet situations to raised mires.

The effect of man can be seen, both in his exploitation of these resources, and in his attempts to tame and restrict these natural changes. Extraction for the water supply results in the flooding of valleys by the damming of rivers and streams, and the raising of the natural water levels of lakes by damming the outlet stream. These reservoirs are used for both direct extraction and for river flow regulation. Energy generation can be seen on the Afon Rheidol where the Central Electricity Generating Board maintain an electricity generating scheme, and historically by the construction of water wheel-driven mills, which abound.

In West Wales streams and rivers are a constant feature of the landscape. Without man's activity ponds and lakes would be much rarer. The map shows the rivers and major tributaries in West Wales. This simplification of the drainage pattern emphasises two features: firstly, a general trend for many of the rivers to have a north/south or northwest/southeast orientation; secondly, the major rivers are often at right angles to this pattern. The initial north/south pattern is thought to have resulted from the landform caused by the 'Alpine' earth movements, when an extensive low-lying level plane was uplifted and tilted to the south. A north/south drainage pattern then developed over this plane. Slowly the properties of the rock strata started to influence this development, with younger, more vigorous rivers developing along lines of weakness in the underlying rock. The most notable of these lines of weakness are the Teifi and Tywi anticlines.

Subsequent earth movements caused by continental drift and sea floor spreading resulted in sea level oscillations and a general long-term lowering of sea level. Long-term trends within these oscillations resulted in the series of wave cut platforms, and considerable uplift in the earth's crust occurred. These changes greatly enhanced the development of the major rivers flowing to the sea, as the drops in sea level rejuvenated them, enabling greater erosion of their valleys and hence continued capture of the older north/south rivers. Within the last 700,000 years the area has been subjected to glaciation. This occurred from both the Irish Sea Ice Cap in a southerly direction and from the Welsh Ice Cap in a south-easterly direction, and has been responsible for the creation of lakes and local alterations of the river system.

These events have created a drainage system comprising elements which provide the full range of habitat types, as well as some rivers whose characteristics are unique within the British context. The small sluggish rills and flushes from springs, and seepages of the moorlands and heaths which abound around Mynydd Preseli, contrast with torrential streams characteristic of the head waters of most of the main rivers. The most common element is the small fast-flowing cobble-bottomed tributary river. Only the Tywi below Landeilo develops into the large lowland river type, gently flowing along its meandering course in a wide, flat floodplain.

Relics of old water mills abound. Their associated works, diversion and feed channels and mill ponds etc provide valuable wildlife habitat. To provide the sudden drop in water channel profile required for the water wheel, leats often of several miles' length were constructed upstream of the mill. Of a slighter gradient than the natural stream, the water flow is often deeper and slower than would have naturally occured, so these leats provide a stream type that would otherwise not occur so commonly.

Within these broad groups, the habitat range is diverse, associated with features such as stream bed boulders, waterfalls, rockpools, shingle and gravel beds, earth banks, muddy floors, fringing swamp and overhanging trees. The effect of the river upon the surrounding land also provides important habitats like fallen trees and bare ground caused by bank erosion, both important for invertebrates and birds.

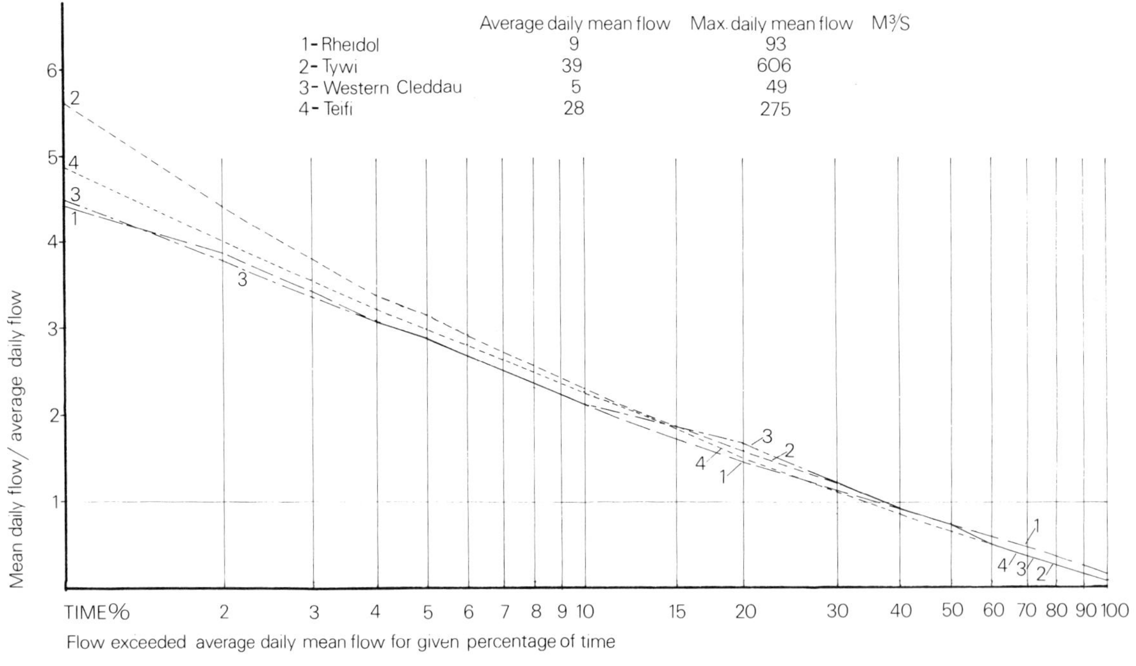

River flows.

Many features of the river are dictated by the flows which vary considerably from river to river. The frequency with which flows deviate from their baseflow, a river's 'flashiness', dictates its stability. A flashy river may cut and fill its banks several times a year and thus adjust its capacity to a flow that occurs relatively frequently, whereas a baseflow river may erode only in rare floods and preserve its enlarged form over intervening years or decades.

Fish of freshwater habitats in Dyfed

	Introduced	Migratory	Non-migratory	Breeding	River	Lake
Sea Lamprey	-	*	-	*	*	-
River Lamprey	-	*	-	*	*	-
Brook Lamprey	-	-	*	*	*	-
Sturgeon	-	*	-	-	*	-
Twaite Shad or Scatsyn	-	*	-	*	*	-
Atlantic Salmon	-	*	-	*	*	-
Sea Trout or Sewin	-	*	-	*	*	-
Brown Trout	-	-	*	*	*	*
Rainbow Trout	*	-	-	*	-	*
Grayling	*	-	*	*	*	*
Pike	*	-	*	*	*	*
Carp	*	-	*	*	-	*
Crucian carp	*	-	*	*	-	*
Gudgeon	*	-	*	*	*	*
Tench	*	-	*	*	-	*
Bream	*	-	*	*	-	*
Minnow	-	-	*	*	*	-
Roach	*	-	*	*	-	*
Rudd	*	-	*	*	-	*
Stone Loach	-	-	*	*	*	-
Eel	-	*	-	-	*	*
Three-spined Stickleback	-	*	*	*	*	*
Ten-spined Stickleback	-	-	*	*	-	*
Bass	-	*	-	-	*	-
Perch	*	-	*	*	-	*
Thick-lipped Mullet	-	*	-	-	*	-
Bullhead	-	-	*	*	*	-
Flounder	-	*	-	-	*	-

The table above shows the species of fish present in our rivers. Interest in salmon and trout fishing ensures that the rivers are managed to maintain healthy populations of these fish and they are widespread. The lampreys and three-spined stickleback are probably present in the lower reaches of all rivers, while the thwaite shad occurs in late spring in the lower reaches of the Tywi and Taf. The sturgeon is only rarely seen in the lower parts of the Tywi and Teifi. The eel is perhaps the commonest and most widespread of fish. The bass, mullet and flounder are primarily marine, but frequently enter fresh water, and indeed the flounder has been found as far upstream as Landovery on the Tywi. The bullhead and minnow are again widespread, but the stone loach is not thought to occur in the northern rivers. Of the coarse fish found in Dyfed, only the pike and gudgeon inhabit flowing water, and these are only found in the lower reaches of the Tywi.

The figure overleaf shows the longitudinal sections of four of our rivers; the Rheidol a steep upland river, the Teifi our longest river, the Tywi our only river developing lowland characteristics, and the Western Cleddau. This selection encompasses most of the features to be noted in West Wales rivers.

The profile of the Rheidol, together with its chief tributaries, shows a steep but variable gradient; it falls some 540m in about 45 kilometres. A large part of its catchment is over 300m with heavy rainfall. The river has a history of frequent and violent floods, and is acid. Mining during the 18th and 19th centuries has left a legacy of heavy metal pollution, although this is affecting the life of the river to a lesser extent now than in the past. Flows in the lower reaches have been modified by the

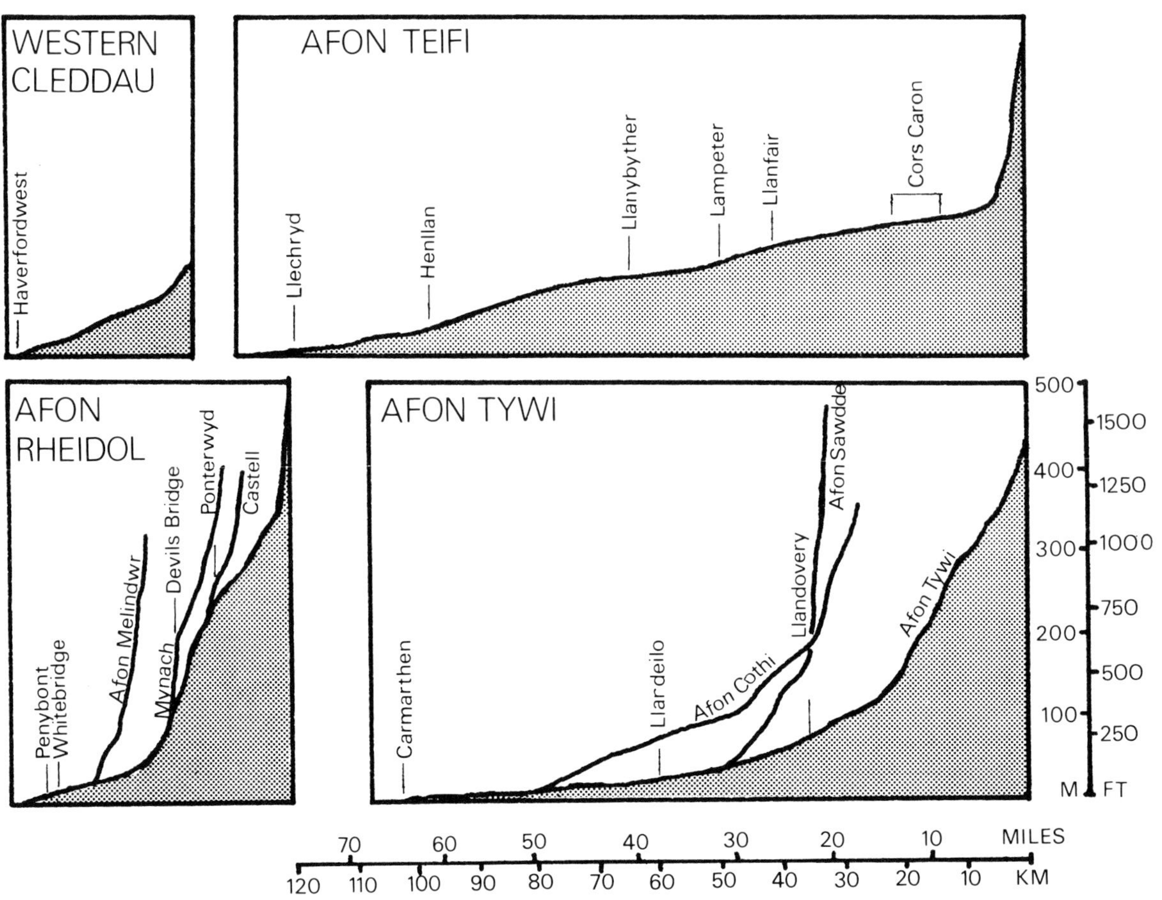

Longitudinal sections of four main rivers.

two upstream reservoirs of Nant-y-Moch and Dinas, which supply water via a tunnel for hydroelectric generation at a rate of up to 15m^3/s. The main characteristic of this river is its strong current, which is continually shifting its bed, and a variable flow. On the lesser gradients the bed is composed variously of flat stones, pebbles and coarse gravel. On its steeper sections, cascades develop, where a series of long, deep, tarn-like pools, scooped out of the rock, are connected by shallow rapids. At normal discharges these pools become quite still and are bottomed by stone and sandy debris. Between storm floods they support great numbers of invertebrates, mainly stoneflies, mayflies and true flies, but the habitat is precarious, only a few surviving the flood. After passing over two waterfalls, its valley opens out and over this section (its lower reaches) it is shallow and erratic in course with great flood beaches of rubble and silt. The figure shows how the course has changed, often leaving pools and banks, or backwater behind. Flowering plants are not able to establish themselves on the shifting bottom, and only where the river is confined along its lower reaches do patches of common water crowfoot and water starwort grow. For the most part mosses and algae constitute the river's flora. Over 60 species of algae have been recorded.

The Teifi, about 140 kilometres long, is our longest river, its source the Teifi Pools, a group of six lakes at about 400m - 450m. It plunges down a river bed made up of solid rock and large boulders. In the Strata Florida valley it slackens, passing over a bed of loose stones, pebbles and gravel until

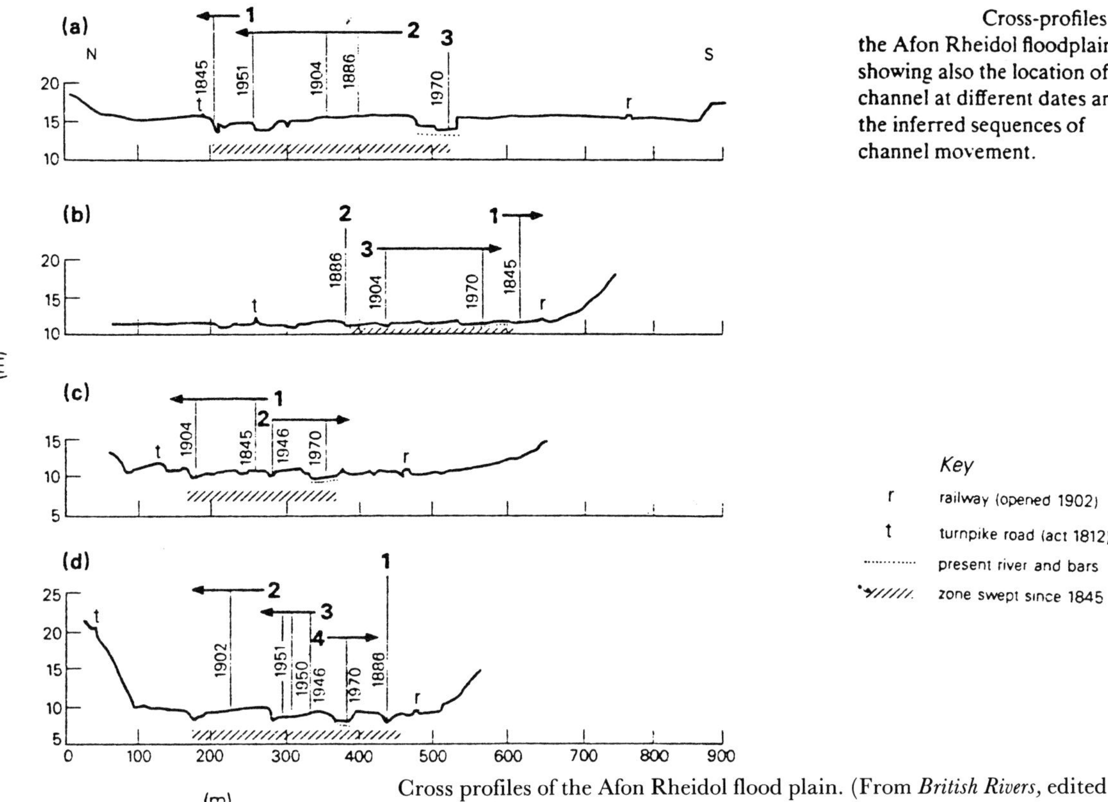

Cross-profiles of the Afon Rheidol floodplain, showing also the location of the channel at different dates and the inferred sequences of channel movement.

Cross profiles of the Afon Rheidol flood plain. (From *British Rivers,* edited by J. Lewin, published by George Allen and Unwin).

at about 12 kilometres it passes into Cors Caron. Here the gradient is so flat that the river wanders sluggishly through the vast quagmire. When not in flood the flow of water can scarcely be detected, and the river consists of a long series of still pools, connected by short lengths of shallower water. A recent survey showed that this section of the river supported at least 100 truly aquatic higher plants, a much higher total in its upper section than any other sections of other rivers rising at high altitudes. This is an indication of the unique character of the Teifi as it meanders through Cors Caron. Below this point the gradient is fairly constant, the river maintains a steady current and is characterised by great beds of water crowfoots and starworts. About two thirds of the way to the sea its bed again steepens and is rocky and irregular. There are small falls at Henllan and Cenarth. Below Llechryd it passes into an imposing gorge about 5 kilometres long before it passes Cardigan and so to the estuary and the sea.

The lower, as well as the higher reaches are much modified by glacial activity. All three features, the gorge and two falls, are on a modified course. It formerly flowed along a completely different valley at these points, this former valley now filled with superficial deposits, its rock floor being at -12m O.D. at Cenarth and -32m O.D. at Cardigan. Surveys on this river have identified at least 147 taxa of macro-invertebrate.

The Tywi is the largest of Dyfed's rivers and is considered to be the nineteenth largest British river. The longitudinal section shows it to have a typical profile of a gradually decreasing gradient as it flows to the sea. Its initial higher section is a narrow steep sided valley, although some of this has now been flooded by the construction of the Llyn Brianne Reservoir. Over the mid-section, the river bed of rubble and silt becomes more erratic until, at Llandeilo, meanders begin to appear. This, the lower section, contains many meanders and the presence of many ox bow lakes shows that the river frequently alters its course. The finest of these ox bows are at Castle Woods and the

Bishop's Pond at Abergwilli. The formation of Bishop's Pond is thought to have resulted when the river changed to a new course during a great flood in July 1802. The flood plain is up to a mile wide and the wet flooded meadows support large numbers of wildfowl during the winter, including the third largest flock of Siberian whitefront geese in Britain. It is along this lower length that the pike and gudgeon can be found, and it is here that the rare and elusive club-tailed dragonfly breeds. The depth of the river is variable, with pools often at deeply cut bends, with a thick silt substrate alternating with gravel banks and cobble-bottomed shallows.

Historically the pH of the water in the river along most of its length has been neutral, perhaps enhanced by the waters of the Sawdde, one of its major tributaries. This river rises in the limestone outcrop and consequently has water of a high pH. There is now some concern that the large forests of the uplands and the effect of the Llyn Brianne reservoir may tend to acidify these waters. Over 130 species of algae and 106 taxa of macro invertebrates have been recorded on the main river. These animals occurred at densities of 1,300 to 4,300 per square metre although prolonged flooding could be most destructive, reducing populations to 160 to 230 per square metre.

The character of the Western Cleddau results from a landscape extensively moulded by glacial meltwater, flowing often under hydrostatic pressure beneath the ice, eroding the bedrock, resulting in channels and hollows which defy gravity. Such meltwater channels, with narrow flat bottoms, are a feature of the North Pembrokeshire landscape. The river rises at about 100m some 3 kilometres from the sea and then flows slowly inland over soft substrate in a wide marshy valley occasionally bordered by extensive fens. On passing downstream, it flows faster over rocks, pebbles and gravel, particularly through Treffgarne gorge. Much of the upper section is wooded. After the gorge, it slows and meanders through a 0.25 kilometre wide, flat-bottomed valley, to the Milford Haven inlet. Many of the higher tributaries flow in old meltwater channels; some rise on moorland, while others are slow-flowing, silty, lowland water-courses. Others remain in an unimproved state and retain a natural hinterland, allowing the catchment to be a major haunt of the otter in England and Wales.

Standing water occurs either where the drainage is impeded by glacial deposition like Talley Lakes, by the build-up of peat in mire systems like Cors Goch just west of Carmarthen, movements of sand in dune systems like the Witchett Pond, Pendine, where water accumulates in kettleholes like Pencarreg or the Mathry Pools and other depressions such as the relic pingoes, or where man has created artificial dams or barriers of various kinds, and where mineral extraction occurs below the water table like the Ludchurch quarries. The table lists the lakes of Dyfed.

The two outstanding features of the distribution of lakes in Dyfed are the myriad small lakes over lowland Pembrokeshire and the concentration of large lakes in the eastern highlands. Surprisingly, the small lakes of Pembrokeshire are a recent man-made phenomena, mainly farm irrigation reservoirs. Their contribution to nature conservation is already becoming manifest, and it is to be hoped that, as they mature, they will provide a mosaic of wildlife refuges across the countryside. They are usually less than 5m deep but vary in area, from small pools not shown on the map, to lakes of up to a hectare. The highland lakes are either purpose-built reservoirs, created by flooding valleys, by damming rivers, or lakes of natural origin, but sometimes modified for use as reservoirs.

The interior of the St David's peninsula is a plateau 40-70m in altitude, with many poorly drained saucer-shaped basins. The centres of the major depressions of Dowrog, Trefeiddan and Waun Llanrduidion have pools covering several hectares, but only about 1.5m deep. Set in common land, grazing has restricted tree growth, but plants such as water horse-tail, bulrush, bog bean, branched bur-reed and marsh cinquefoil are reducing both the extent and depth of the open water.

Lakes of Dyfed

	O.S. Grid ref.	Approx altitude m	Approx area ha	Capacity $m^3 \times 10^6$	Water Extraction/ regulation	Comment
Bosherton	SR9794	4	33	-	-	Marl lake,NNR
Claerwen Reservoir	SN8365	365	270	49	+	Deep
Cwm Rheidol	SN7079	45	24	0.68	+	Dammed valley CEGB in 1950's
Dinas Reservoir	SN7482	260	25	0.99		Dammed valley CEGB in 1950's
Falcondale	SN5649	140	4	-	-	Dammed, over 100 years old, shallow
Lliedi - lower	SN5103	70	14	0.82	+	Dammed valley, built 1878
- upper	SN5104	76	14	0.82	+	Dammed valley built 1902
Llyn Berwyn	SN7456	440	15	-	-	Shallow, oligotrophic
Llyn Brianne	SN8050	300	216	61	+	Deep, valley built in 1972
Llyn Blaenmelindwr	SN7183	320	6.5	-	-	
Llyn Conach	SN7393	410	12	-	-	
Llyn Craigypistyll	SN7286	320	11	0.35	+	
Llyn Du	SN7969	540	2	-	-	
Llyn Dwin	SN7292	410	2	-	-	
Llyn Eiddwen	SN6066	300	10.5	-	-	Shallow, SSSI WWTNC reserve
Llyn Egnant	SN7967	430	18	0.52	+	Natural lake level raised by dam in 1965
Llyn Fanod	SN6064	315	4.5	-	-	WWTNC reserve
Llyn Frongoch	SN7275	275	7	-	-	
Llyn Fyrddon-Fach	SN7970	540	4	-	-	
Llyn Fyrddon-Fawr	SN8071	535	13	-	-	SSSI
Llyn Glandwgan	SN7075	275	7	-	-	
Llyn Gwaith	SN6750	426	4	-	-	
Llyn Gwngu	SN8372	440	3	-	-	
Llyn Gynon	SN7965	430	25	-	-	SSSI
Llyn Hir	SN7867	450	7	-	-	
Llyn Isaf	SN8075	490	2	-	-	
Llynoedd Ieuan	SN7981	545	5	-	-	
Llyn Llech Owen	SN5615	240	4	0.06	+	No dam, shallow
Llyn Llygad Rheidol	SN7987	530	6	0.30	+	Morraine dammed lake
Llyn Pendam	SN7083	320	3	-	-	
Llyn Pen-rhaidhr	SN7593	410	7	-	-	
Llyn Plas-y-mynydd	SN 7492	390	6	-	-	
Llyn Rhosgoch	SN7183	335	3	-	-	
Llyn Rhosrhydd	SN7076	275	6	-	-	
Llyn Syfydrin	SN7284	330	13	-	-	
Llyn Teifi	SN7867	410	25	0.70	+	Level increased by dam in 1960
Llyn y Fan Fach	SN8021	520	11	0.91	+	Deep

	O.S. Grid ref.	Approx altitude m	Approx area ha	Capacity $m^3 \times 10^6$	Water Extraction/ regulation	Comment
Llyn y Gorlan	SN7866	450	3	-	-	
Llys y Fran	SN0324	105	78	9.11	+	Dammed valley built 1971, deep
Llywernog	SN7281	300	4.5	-	-	
Nant-y-Moch Reservoir	SN7586	340	270	32.55	+	Dammed valley, CEGB in 1950s
Orielton Lakes	SR9599	30	4.5	-	-	Dammed valley
Pencarreg	SN5345	100	9	-	-	Kettle hole, SSSI oligotrophic
Pond yr Oerfa	SN7279	330	4.5	-	-	
Rosebush	SN0629	230	13	0.64	+	Dammed valley built 1913
Talley Lakes	SN6333	100	18	-	-	Formed in glacial hollows - SSSI
Usk Reservoir	SN8328	320	121	12.27	+	Built in 1955, deep
Witchett Pool	SN2087	6	9	-	-	SSSI

At Bosherston in South Pembrokeshire the lake occupies a drowned valley system only a few metres above sea level in the carboniferous limestone. The sea has been progressively kept out of the valley, until about 130 years ago the outlet gorge was blocked, enclosing a lake of about 33 hectares in three fingers, but with a maximum depth of only 3.5m. Although enrichment by agricultural run-off is a problem its central finger still exhibits classic marl lake conditions, for which the lake system is nationally important. Here large beds of stonewort *Chara* species still thrive. Elsewhere the white water lily is an attractive feature but, in the shallow waters, they totally enshroud the surface and shade out other plants.

Within the 9 kilometres dune system of Pendine, developed on the western side of Carmarthen Bay, lies an extensive marsh, resulting from encroachment of sand dunes blocking drainage from the inland plateau. Within this marsh is a shallow, sandy bottomed 9 hectare calcareous fresh water lake known as the Witchett Pool. This lake, together with other open water within the marsh, supports greater bladderwort, stoneworts *Chara* species, watermilfoils *Myriophyllum* species, pondweeds *Potamogeton* species and soft hornwort. A reed swamp surrounds the pool.

The Upper and Lower Talley lakes are natural pools at about 100m altitude, extending to about 18 hectares to occupy hollows in glacial deposits. The shallow lower lake has retained a complete vegetational sequence from open water, through reedswamp, to mature alder carr. To the south the upper lake is more open and supports blunt-leaved pondweed; water sedge is also found. Breeding birds include great crested grebe, while the lakes in winter support a good range of duck.

The lake Llyn Eiddwen is some 10.5 hectares in area with a maximum depth of about 4m, sited in a wide, virtually treeless valley at about 300m altitude on the Mynydd Bach. The outlet to the lake is via extensive beds of bottle sedge and water horsetail, which grades into wide peat bog. Water lobelia, shoreweed-quillworts, awlworth and floating water plantain may be found in the open water, an almost unique aquatic community.

The goosander is extending its range in West Wales. (DF)

The largest reservoir in Dyfed is Llyn Brianne and it may be considered typical of recent purpose-built reservoirs,.The rockfill dam that blocks the course of the Tywi rises about 80m above the old river bed. There are a few shallow areas, most banks falling steeply, and the maximum depth must be about 75m, while the average is about 28. Wave action maintains an eroding shoreline hostile to both plants and animals, and fluctuations in water level exacerbate the situation. Because of the steeply sloping shore, no sub-littoral communities have become established. The depth does not allow nutrient recycling, as water movements, due to both thermal currents and wave activity, are confined to the surface layer. Nutrients that enter the lake are locked in the bottom mud. Primary productivity is low, being restricted to phytoplankton in the surface zone of the lake. It is here that light penetration is sufficient to allow growth but, because of the lack of nutrients, even the activity of this community is low. The lake has little natural history interest.

Otter

ABOVE: Llyn-y-Fan-Fach below Bannau Sir Gaer. (BBNP) BELOW: Bishop's Pond, an ox-bow lake near Carmarthen. (NCC)

ABOVE: Ludchurch quarry pool. (JWD) BELOW: Pool at Lower Treginnis on the St David's peninsula; many of the smaller ponds are especially important for plants and invertebrates. (SBE)

ABOVE: The upper Egnant below Llyn Egnant. (JPS) BELOW: The Teifi at the south end of Cors Caron, showing the rich aquatic flora of this slow moving river. (JPS)

ABOVE: Excavating new pools at the Dowrog Common Nature Reserve, aided by the Welsh Water Authority. (DRS) BELOW: One of the pools, the result of the above work. (JWD)

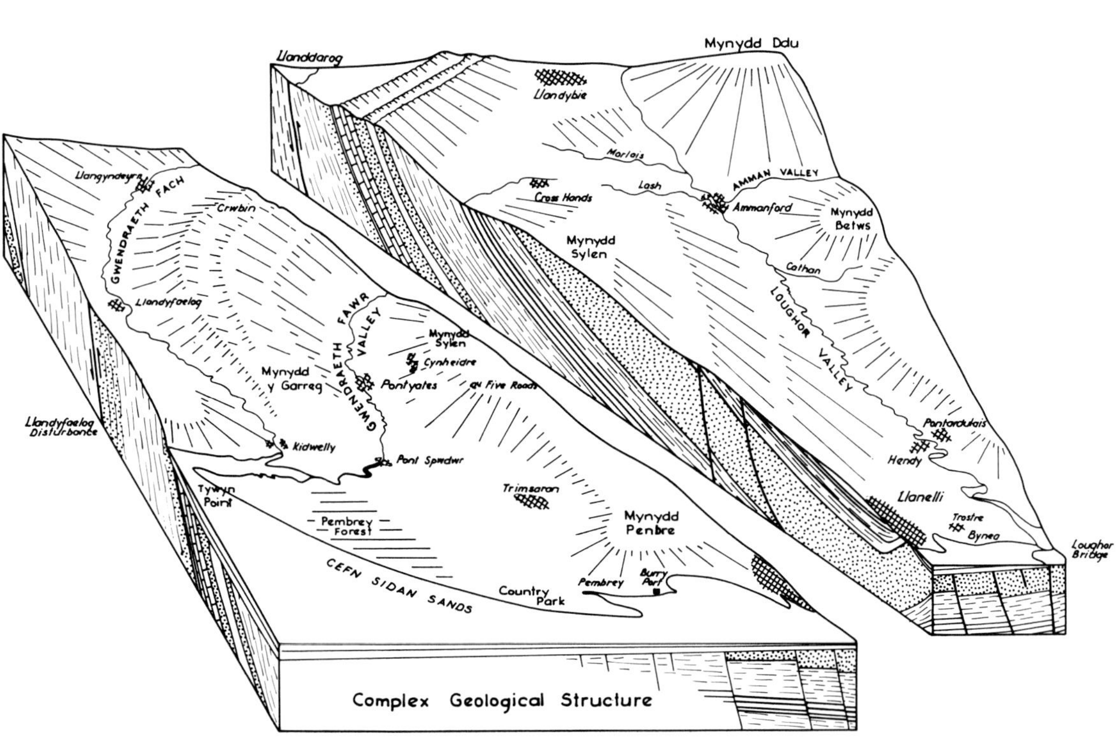

Geological Structure of south-east Carmarthenshire.

The Industrial South-East

Remains of Pencoed Lead House, Llanelli

Much of south-east Carmarthenshire is rural and it also contains a great diversity of natural and semi-natural habitat, but the area mainly comprises the coalfield.

Much of the narrow, ribbon-like, outcrop of the Carboniferous Limestone has been quarried in the past, and there are several large, active workings now, principally supplying roadstone and aggregates. There are also small areas of largely untouched, native limestone woodland between Meinciau and Llandybie, supporting a rich flora. Ash and hazel are the dominant species, with frequent holly, spindle, guelder rose and wild privet, traveller's joy often festooning the trees. Typically the ground-flora is dominated by dog's mercury or ramsons with pignut, red campion, sweet woodruff, early dog violet and early-purple orchid, and also occasionally includes herb paris, lily-of-the-valley and toothwort.

The floors of the abandoned quarries usually support an ephemeral flora of species such as rue-leaved saxifrage, hairy rock-cress and common whitlow-grass. Rusty-back fern and bladder campion can be found on the rock ledges, while spring sedge, hairy oat and yellow oat occasionally occur in grassy areas. Pools in some old workings provide an ideal habitat for stoneworts. Some quarries have been disused long enough for woodland to have grown up, and their flora may approach that of the native limestone woods. Notable invertebrates also occur, including the bee chafer, wasp beetle and small blue and marbled white butterflies.

East of Llandybie the limestone outcrop forms part of Mynydd Du and consists largely of heavily grazed moorland. Only inaccessible crags, many of which are the result of former quarrying, support a recognisable, ungrazed flora. Here mossy saxifrage, carline thistle, wild thyme, common butterwort and green spleenwort struggle against the sheep.

The Millstone Grit outcrops in a narrow band, which widens somewhat to the east, on the Mynydd Du and, in simple terms, consists of two major quartzitic sandstone units separated by a softer shale sequence. West of Llandybie, the lower sandstone, the Basal Grits, a pure quartzitic rock, has been quarried for refractory brick and glass making and for roadstone, and forms a prominent escarpment. The acid dip-slope, often devoid of soil, is an environment hostile to most plant life.

The intervening Middle Shales erode to a waterlogged hollow and in places basin mires have developed. Species often encountered include cross-leaved heath, sundew, bog asphodel, royal fern and occasionally white beak-sedge and bog myrtle. The dew-pond of Llyn Llech Owen, now a drinking-water source, is also situated in the same depression. Several scarce aquatic plants occur here, including both white and yellow waterlilies. The site is also notable for its dragonflies and supports a substantial colony of the black darter, one of six populations known in Carmarthenshire.

The upper sandstone member, known as the Farewell Rock because the old miners knew that no coal would be found below it, forms a less well defined escarpment and contains some base-rich horizons, especially on the northern slopes of the lower Gwendraeth Fawr valley. Near Glyn Abbey, several ravines are deeply incised into the dip-slope and are well wooded and rich in plants which thrive in damp, humid conditions.

The Gwendraeth Fawr and Amman valleys are cut in the more easily eroded Lower Coal Measures. These rocks are often overlain by thick deposits of glacial boulder-clay and, as a result, the soils are acid and poor-draining. The traditional practice of miners having a few cattle-rearing acres to supplement income has meant the land has never been intensively farmed, and the use of agrochemicals and modern drainage techniques are rare. Characteristic habitats include wet, species-rich pastures and semi-natural woodlands.

The pastures are usually horse or cattle-grazed, damp, heathy and often undulating: the result of disturbance from old, shallow mine workings. The flora invariably includes whorled caraway, a characteristic plant of the Carmarthenshire coalfield. This species seems to tolerate a fair spectrum of pH and wetness in this county and it is not fully understood why it should be so rare elsewhere. In the drier, heathy areas, sometimes the result of low 'tumps' of old, grassed mine-spoil, the caraway or *Carum* may be associated with heather, tormentil and yellow rattle, while in moderately wet environments, typical associates are meadow thistle, devil's bit, saw-wort, petty whin, marsh orchids and lousewort. The wettest sites often include bogbean, marsh valerian and lesser skullcap.

The coalfield woodlands are mostly confined to steep valley slopes, or areas abandoned due to the preponderance of dangerous old pits, but some of the oldest tips are also well wooded. They comprise dominant oak with ash, hazel, and frequent holly. Sanicle and moschatel are often constitutents of the ground-flora, but in wetter situations wood horsetail and dog violet may be common. Woodland birds are abundant, with many pairs of wood warbler and tree pipit, and high densities of buzzard. In recent years the pied flycatcher has become established as a breeding species.

Many of the younger colliery spoil tips support a rich flora and fauna, but are increasingly subject to reclamation schemes. Higher plants recorded on the Great Mountain Colliery tips at Tumble exceeded 180 species in 1974 and included a large and varied population of marsh orchids, abundant wood horsetail and common broomrape and kidney vetch, a plant usually only found on the Carmarthenshire coast. Common water plantain, greater pond sedge and bulrush grew in silted lagoons. Butterflies were particularly abundant: small copper, common blue and small skipper, all often seen resting on the warm, sunny, black shaly slopes. The site has now been reclaimed.

The modern practice of open-casting shallow coal reserves has been responsible for the loss of much semi-natural habitat in the area, but there appears to be an increasing awareness of this environmental damage by the authorities involved. The massifs of Mynydd Penbre, Mynydd Sylen and Mynydd Betws are formed from the hard, resistant, Pennant Sandstone, which overlies the main coal-bearing rocks. The scarp-face of the formation is responsible for the steep southern slopes of the Gwendraeth and Amman valleys. The rivers Lliedi and Morlais rise in the Cynheidre-Llannon area in extensive, acid basin mires, the flora and fauna of which may be rich, often having a *Carum* meadow-type association, but in addition including sundew, royal fern, cranberry, bog pimpernell, marsh St John's wort and cotton grass when conditions are favourable. In a few of the best sites, particularly around Llannon, extensive areas of well grown bog myrtle, royal fern colonies reaching 2m in height and several stands of narrow buckler fern occur. It is presumably hereabouts that James Motley recorded marsh gentian in about 1850, a species not seen in the county since, despite repeated searching. Colonies of marbled white and marsh fritillary butterflies have been recorded in these undrained semi-natural habitats, but concern has been expressed recently at the destruction of some sites by peat-cutting.

LEFT: Herb paris still occurs in some limestone woodlands in the south-east. (DF) RIGHT: Marbled white butterflies are found in both the whorled caraway meadows and in the dune areas of south-east Carmarthenshire. (DF)

The two Lliedi Reservoirs to the north of Llanelli do not support a rich flora, but their avifauna is of interest. A pair of great crested grebes have attempted to nest annually in recent years, meeting with varied results, being more successful in wet years when reservoir levels tend to remain constant. Pochard and tufted duck are resident for much of the year and ospreys are seen on passage in some autumns. Rare winter visitors have included scaup, Bewick's swan, ruddy duck and glaucous gull.

The coastline of south-east Carmarthenshire is largely one of accreting sediments. Cliffs at Pembrey, reputedly washed by the sea in Roman times, are now degraded and protected by large areas of duneland, saltmarsh and mudflats. The six mile, sandy beach of Cefn Sidan fronts the extensive dune system of Tywyn and Pembrey Burrows. In the lee of the dune ridges, saltmarshes have built up which, over the centuries, man has reclaimed for agriculture. More recent changes have been the planting of Pembrey Forest, the construction of Pembrey Airfield and the development of Pembrey Country Park. Before the Swan Pool was drained during airfield building, flocks of 400-500 white-fronted geese were seen in most winters, but today the species only roosts in the area on the coastal mudflats, and diurnally migrates to its feeding-grounds in the Vale of Tywi. The intertidal sands of Cefn Sidan are an important British site for sanderling, where up to 1,000 birds may be seen in May.

Familiar strandline plants are found above the high water mark, including sea rocket, sea sandwort and several sand-binding grasses. The fauna is more diverse and the rare strandline beetle *Eurynebria complanata* has been recorded. Drift-wood washed up on the beaches, particuarly at Tywyn Point, provides shelter for small animals, and the scarce, small, white pillbug, *Armadillidium album*, toads and common newts are often found hiding beneath such debris.

The dry dune ridges closest to the sea are unstable, and present a hostile environment in which only well adapted plants like marram grass can grow. As the dunes mature, plants such as seaside pansy and pyramidal orchid become established, and one dry area of Tywyn Burrows has a healthy colony of bee orchids. Sea buckthorn was introduced to assist stablization when the forest was planted in the 1930s, and now completely dominates much of these exposed dunes, to the detriment of the native flora, although birds feast on the berries in winter.

The hollows between the dune ridges, termed 'slacks', always damp and usually containing standing water in winter, have the richest flora. One of the county's specialities is fen orchid, not seen on Tywyn Burrows however since 1971, and on Pembrey Burrows since c1930, but still present in the county on Laugharne Burrows. Confined in western Britain to Carmarthenshire and Glamorgan, this species has suffered in recent years at Tywyn Burrows from the loss of suitable habitat, due to the lack of grazing, a factor that has allowed low scrub of creeping willow to become dominant and smother out the delicate species. Another rare plant similarly affected is the dune gentian, confined to five sites in south-west Wales but still seen in most years on Tywyn Burrows.

To the non-naturalist, saltmarshes appear to be monotonous, dull places and in the past, planners invariably seemed to recognise them as ideal land for industrial development. The Llanelli area has suffered more than its fair share of such despoliation, but vast areas of estuarine habitat are still available to wildlife. The saltmarshes at Pembrey and Tywyn Point contain a large proportion of sand in their make-up, a feature unusual in western Britain, and their wide floral zonation and rapid accretion make them scientifically important in a national context. Sympathetic development is now proposed to the east of Llanelli, where a wildfowl park and refuge is to be constructed on the saltings, as part of a plan to rejuvenate the town's derelict coastal industrial sites.

The coastal burrows support a varied fauna in addition to their unique flora. Butterflies of the dunes and marshes include marbled white, small blue, silver-washed, dark-green and marsh fritillaries, grayling and brown argus, and in one forest ride thirty-two species have been recorded. The avifauna is no less notable. Raptors regularly seen hunting over the wide, open areas are peregrine, merlin, hen harrier and short-eared owl. The tidal muds provide rich feeding grounds for waders, species likely to be seen at any season being oystercatcher, curlew, redshank and ringed plover. Important winter populations of dunlin, knot, bar-tailed godwit and grey plover are present and passage species include black-tailed godwit and ruff. The large Carmarthen Bay flock of common scoter is sometimes visible offshore under favourable weather conditions, and other duck common in the Burry and Gwendraeth estuaries include shelduck, pintail, eider and red-breasted merganser.

Freshwater wetland is another important habitat near the coast and the WWTNC reserve at Ffrwd Farm, Pembrey, contains some species-rich fen communities. Foremost among the plants to be found is a strong colony of marsh pea, first discovered in 1971 and here at one of only two Welsh sites. Other species growing here are frog-bit, tubular water-dropwort and floating club-rush. The fauna is no less spectacular and several dragonflies have been recorded, including the hairy dragonfly, a species suffering decline in Britain as a whole. Another invertebrate, frequent on the reserve, is the short-winged conehead, a type of bush-cricket. Reed buntings, reed and sedge warblers nest in the tall fen, and water voles have been seen in the pills. In the winter months, a variety of wildfowl and waders are seen, but it is also proposed to develop the reserve to attract a greater diversity of summer birds.

One legacy of Llanelli's industrial past is the large area of disturbed and abandoned land on the immediate coastal strip. Relict natural habitat remains but is usually rather decrepit. The Machynys Ponds, for instance, despite the rubbish tipping and adjacent factory shells, provide a suitable environment for wildfowl, and their flora is rich, including the only Carmarthenshire site for ivy-leaved duckweed. Old Castle Pond, in the centre of the town, a disused industrial reservoir, has an abundant flora and fauna and attracts wildlife for the townspeople to see at first hand. Of more specialised interest are the several species of rare hoverflies and dragonflies to be found here, together with the short-winged conehead, common amongst the rushes.

Opportunist plants are much in evidence: hoary mustard is an alien species which has become established on any suitable bare ground, and Oxford ragwort is a common street weed. The old canals of the lower Gwendraeth valley have been colonised by many of the plants of the fenland communities and great water dock. The railways which displaced the canals, now themselves disused, also have their unique plants: most notable is prostrate toadflax, a species established here but only casual elsewhere in Wales, which is often associated with rat's-tail fescue.

Lily of the Valley

ABOVE: Crwbin Common, an area of rough pasture on the edge of the coalfield. (RP) BELOW: Prostrate toadflax beside a disused railway line. (PP)

ABOVE: Embryo dune system near Tywyn Point. (SBE) BELOW: Sea buckthorn encroaching on a new dune slack near Tywyn Point. (SBE)

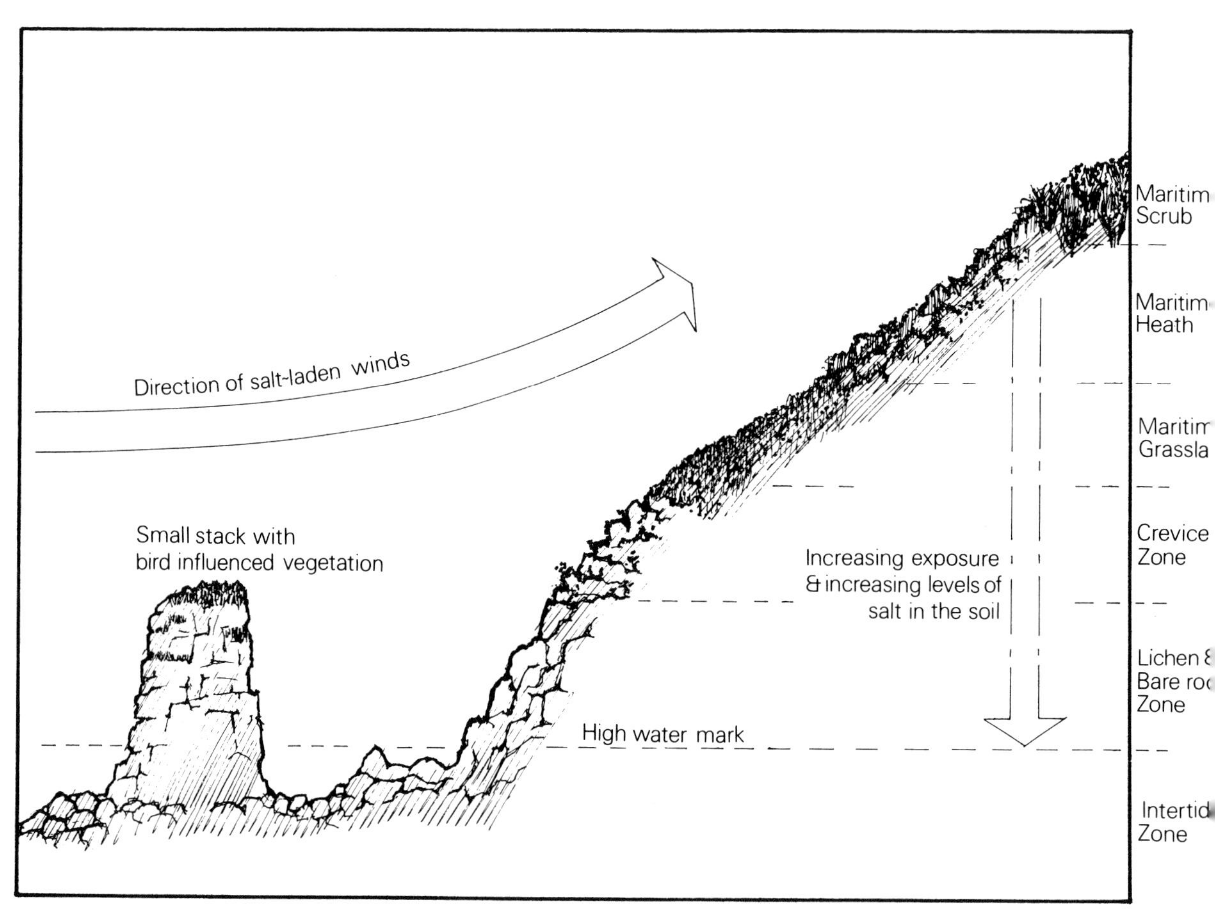

Cliff Vegetation Sequence.

The Mainland Coast

Chough over Marloes

The coast of West Wales has an enormous concentration of rich and varied wildlife habitats. The coastline extends for some 350 kilometres and very little is subject to urban or industrial development. Although coastline is, by definition, a narrow feature, the major estuaries of Carmarthen Bay and the Dyfi have extensive inter-tidal flats, salt marsh and sand dune. Away from the more sheltered coasts, the splendid rocky headlands and cliffs of the far west have achieved national recognition as the Pembrokeshire Coast National Park — the only truly maritime National Park in England and Wales. It is here that the coastal water of Cardigan Bay and the outer Bristol Channel meet the warm Atlantic waters carried by the Gulf Stream, and here also that resistant Carboniferous Limestone and intrusive igneous rock form spectacular headlands and islands facing the all-too-frequent fury of the Atlantic storms.

Many of these cliffs date back before the last Ice Age and have been battered by the waves of inter-glacial seas. The presence of raised beaches like the fine example at Poppit, at the mouth of the Teifi, provide evidence of a higher sea-level during previous inter-glacials. Conversely, the submerged forests exposed on numerous sandy beaches such as at Borth, Newport, Whitesands, Newgale and Marros, testify to lower sea levels. These areas of peat with tree bases and trunks of pine, alder, hazel, birch and oak, have been dated back to the warmest period after the last Ice Age, when forest covered much of the country, about 6,000 years ago. Newgale's submerged forest was reported as early as 1188 by Giraldus Cambrensis in his *The Journey through Wales*. He remarked when crossing Newgale Sands on 24 March that, following the stormy winter of 1171-2, 'Tree trunks became visible, standing in the sea with their tops lopped off, and with the cuts made by the axes as clear as if they had been felled only yesterday. The soil was pitch-black, and the wood of the tree trunks shone like ebony'.

The glaciers and ice sheets that moved down the Irish Sea and out from the Welsh mountains left behind vast quantities of clay, sand, stones and boulders. When the ice retreated and sea level rose this debris was re-worked by coastal processes to form the beach material of today. In the shallow waters of Carmarthen Bay, wide expanses of inter-tidal sand, largely of glacial and fluvio-glacial origin, has allowed huge dune systems to develop from the sand blown inland by south-westerly winds. Salt marshes flourished in the lee of the dune ridges. Along much of Dyfed's coastline, the sea has regained its earlier cliff-line. Where deep water close inshore leaves cliffs fronted by narrow beaches, marine erosion can progress unhindered by wide inter-tidal flats. In a few places, as at Aberaeron and Llanrhystyd, extensive soft cliffs of glacial material are still being rapidly eroded, while the former cliff-line lies just inland, mantled by glacial deposits. Elsewhere, a magnificently varied rocky cliff-line, abounding in caves and with offshore stacks of every shape and size, natural arches and blow holes from collapsed caves, is fronted by rocky beaches, boulders, shingle or sand, all of varying colours.

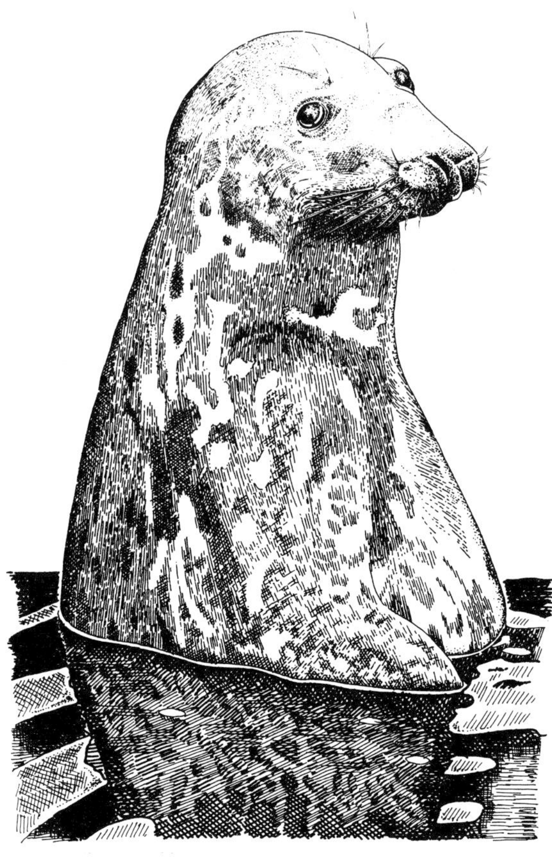

Female grey seal. (DF)

Rocks ranging from pre-Cambrian to Carboniferous in age, and the different structural features resulting from the Caledonian and Variscan mountain building phases, make the coast a geologist's Mecca. The constant marine erosion keeps the majority of cliffs freshly exposed for those undertaking serious research, or for parties of amateurs from the Field Study Centres at Orielton and Dale. In complete contrast is the drowned river system of Milford Haven, where deep, sheltered, saline water extends far inland before being influenced by fresh water from the rivers of the Eastern and Western Cleddau. In 1603, George Owen, aptly described these rivers as 'lovingly joining to make the fair haven of Milford become both a salt sea of a mile broad and 16 miles long before they forsake their native country for whose good they send forth many fair branches on either side, serving diverse towns, villages and gentlemen's houses'.

The majority of visitors to the coast of West Wales are unaware that the landscape of cliffs and valleys extends beneath the sea's surface. Even in the middle reaches of Milford Haven at Lawrenny and near the toll bridge at Warrior Cliff, the rocks descend below low water like drowned river gorges. Concern about the effect that the oil industry of Milford Haven could have on the marine environment stimulated the establishment of the Field Studies Council's Oil Pollution Research Unit at Orielton in 1967. Their diving marine scientists have done much to explore the hitherto secret rocky underwater world of West Wales. They found fascinating plants, and animals of bizarre shapes and colours, occupying an enormously varied environment. This diversity is a response to wide differences in sub-stratum and topography and to a great range of exposure to tidal streams and waves. Outstanding rocky sublittoral areas include flat limestone bedrock off Stackpole Head, the world of Ramsey Sound with its depths of 50m and its narrows with tidal streams of 6 knots, bearing comparison with parts of the Menai Straits in North Wales, and the extremely exposed locations around the 24 islets of the Bishops and Clerks and around Grassholm. Abereiddy quarry is the complete opposite, in that extreme shelter from wave action and tidal streams has been created in a fully saline situation and is unique in south-west Britain. The waters around Skomer encapsulate much of the variety of the underwater environment of West Wales, and it is no coincidence that this area became the first Welsh voluntary marine reserve in 1976.

In contrast to the private world of the diver, the delights of rock pools, nature's infinitely varied marine aquaria, and life between the tides, are available to all. The quirks and charms of the foreshore are no secret to those who have benefitted from the inspired field teaching of John Barrett, for many years warden at Dale Fort. In this he has illustrious forebears. In Victorian times, Tenby was a focus for much pioneering interest in the natural history of the seashore, shared by eminent figures like T.H. Huxley and Dr Samuel Wilberforce. As early as 1 September 1856, Philip Henry Gosse was leading seashore classes to Tenby's caves and rocks and to Monkstone Point. The shore classes were supplemented by indoor talks and evening work arranging pupils' private aquaria. His book *Tenby: A Sea-side Holiday* contains fascinating details of shore excursions with family and friends during the summer of 1854. In his own words, 'Nearly every day's occupation is set down, just as it occurred; tide pool exploration, cavern searchings, microscopic examinations, scenery huntings, road-side pryings — here they all are a faithful narrative of how the Author was engaged for about 6 weeks at that pleasant little watering place'. It is perhaps prophetic that by 1856 he was commenting that, since 1854, Tenby's caves and rocks had been considerably hacked about and had lost many of their species, especially sea-anemones, to collectors. Fortunately, such human predation is rare today, and the shore environment remains of the highest quality, with parts ranking among the best examples in Britain. With its richly varied geology, the extensive deep-water sheltered inlet of Milford Haven and the islands of Skomer, Stockholm and Ramsey with their adjacent headlands, present the full range of intertidal marine habitats. They vary from fully exposed to extremely sheltered, within a small area. The extensive studies and publications of the Field Studies Council reveal that their quality remains virtually unaffected by the proximity of one of Britain's largest oil ports.

Away from the bustle of Milford Haven and the sandy beaches of the south, the late visitor to the coast may encounter the grey seal hauled out on remote, rocky beaches or in the many sea caves. Here the cow will give birth to its white-coated pup from early autumn through to winter, or even later. Sporadic late winter and spring births are a particular feature of the grey seal colonies of West Wales. On isolated mainland beaches, haulouts of up to 350 seals occur, where they rest and complete their annual fur moult. Enormous, battle-scarred bulls and veteran cows claim the prime beach positions, while young animals romp in the surf. West Wales'population of grey seals — of about 3,000 — is the most southerly major population of this mammal in Europe, and was vividly described, nearly 400 years ago by George Owen, 'The seal is covered with hair, like a calf, and hath four short legs, and broad pawed like to the mole, this fish cometh to land to rest and sleep, and lie together in hordes, like swine one upon an other, and at birth time, as Pliny says, cometh to land and is delivered and giveth suck to the young, till he be able to swim'.

Towering above the remote seal beaches are cliffs, often up to 100m high. The varying colours of the weathered rock are further mottled by the black, greys, browns, greens and yellows of the maritime lichens. These are the first plants to colonise the rocks bared by marine erosion and, on the more stable cliffs of ancient igneous rocks of Skomer and Strumble Head, and on the old red sandstone at Manorbier, nationally important lichen assemblages occur. These rugged cliffs are the haunts of peregrines and choughs as well as ravens and buzzards, while interestingly a few house martin colonies occur and there are some cliff-nesting swifts. About 60 pairs of chough breed on the West Wales coast, and their numbers appear to be relatively stable, while peregrine numbers are increasing, with about 30 breeding pairs. Although the main seabird colonies lie on the offshore islands, some ledges on mainland cliffs are equally safe from mammalian predators. A total of about 4,000 individual guillemots, 500 individual razorbills and 750 pairs of kittiwakes can be viewed at Stackpole Head, Elegug Stacks and Green Bridge, on the massive Carboniferous Limestone cliffs of the Castlemartin peninsula, now so popular with cliff climbers. There are voluntary constraints on climbers during the breeding season, and in fact the numbers of auks have actually increased in recent years. Further north there are scattered auk colonies with significant concentrations at Needle Rock, Fishguard, and Needle Rock, Dinas, and especially notable colonies south of Newquay. Together, these total some 3,000 guillemots, 500 razorbills and 200 pairs of kittiwakes. Censusing these mainland colonies is not easy, particularly those in Ceredigion, and their full size was not appreciated until seaborne counts were organised in the late 1970s. Other breeding seabirds include the ubiquitous gulls, shags and cormorants. Colonies of the latter are especially notable in Ceredigion, where up to 200 pairs regularly nest at Pen-deri, south of Aberyswyth.

The guano around these seabird colonies provides a nutritious soil for salt tolerant plants. The tree mallow with its showy purple flowers and velvety green leaves is a characteristic plant, along with sea beet. Specimens of the biennial tree mallow often exceed the height of a man by their second year, providing dramatic proof of the abundance of nutrients in the guano that collects in the crevices. Elsewhere, only a few species are tough enough to grow on the seaward faces of the cliffs. Crevices shelter rock and golden samphire, thrift, sea campion, and rock sea lavender. Chives prosper in the crevices of igneous rock at St David's Head, where it flowers profusely. On similar, presumably base-rich igneous rocks near Strumble Head, the beautiful, deep blue spiked speedwell adorns a south-facing rocky slope in July.

The exposure to strong, salt-laden winds gives rise to a regular cliff-top vegetation sequence, from the scattered plants in rock crevices to grassland, heath, scrub or bracken, and then occasionally to wind-pruned oakwood. This maritime grassland occurs all along the cliff coast, even far up into Milford Haven. The blue-greens of red fescue grass are prominent where there are no herbivores and, in their prolonged absence, a dense, springy turf or mattress develops. Grazing of these grasslands benefits the chough by providing accessible feeding grounds, rich in invertebrates, near to its nest sites. Where grazing occurs, thrift flourishes, and the maritime grassland takes on a pink hue in late spring. The cliffs are a blaze of colourful flowers by then and the display continues right through the summer. Few spectacles can be more breathtaking than the grassy slopes near Strumble Head in spring, where the yellow of cowslips often mingles with swarms of early purple orchids, or the south-facing slope of St David's Head in early June, where the yellow heads of kidney vetch seem to dance through the grassland into the maritime heath above. Sky blue sheets of spring squill adorn both grass and heathland. This small, coastal lily, supposedly a strongly maritime plant, is found as far inland as 11 kilometres at Treffgarne Rocks. Its near relative — the bluebell — is frequent in shady, north-facing gullies low down on the cliffs, as well as under the extensive beds of bracken which dominate too many of the sheltered, deeper-soiled slopes.

On shallower, exposed areas, especially in the north on ancient igneous rocks, maritime heath abounds. Heather and bell heather dominate and even occur on the limestone in the south. These heaths share many rarities with the lowland heaths just inland, but some, like the field gentian, are now confined to the cliffs. Another is the parasitic common dodder which has one of its Welsh strongholds west of Porthclais, despite the frequent fires that affect these heaths at the end of most winters. Broom is frequent in the heaths north of Milford Haven, and has adapted to the strong winds by evolving a distinctive prostrate subspecies, *maritimus*. On the east side of Ramsey, a long-isolated population of juniper has made a similar adaption: the four remaining bushes are intermediate between the prostrate and upright subspecies of the rest of Britain. It is probably no coincidence that the junipers with the closest characters come from another relict maritime heath population of a few bushes on the Lizard in Cornwall. This evergreen was, according to pollen

LEFT: Rock sea lavender. RIGHT: Spring squill.

evidence from the peat bogs of West Wales, once widespread in the locality after the retreat of the ice. Only two other ancient coastal bushes are known - one at Cemaes Head and one at Skrinkle Haven - but fortunately cuttings from the Ramsey junipers are growing safely in the botanic gardens at Cambridge.

The presence of Western gorse in the heath marks a transition to the taller, woody growth of more sheltered slopes. On the cliffs between St David's and Strumble Head, the transition is most pronounced, perhaps because of the tradition of burning, and here the resultant riotous mix of yellow gorse and purple heathers enthralls the many August visitors to the Pembrokeshire Coast National Park. The less hardy common gorse is abundant everywhere, and its flowers are almost too rich for walkers in May, when their deep yellow blooms and over-powering perfume assail the senses. In places, a wind-pruned canopy of blackthorn and privet clings to the cliffs, with the addition of dogwood in Carmarthenshire.

Stonechats are found along much of the cliff coastline of West Wales. (DF)

Birds of the scrub and heath include breeding stonechats, wheatears, whitethroats, linnets and yellowhammers. Of the insects, bloody-nosed beetles are common, while the fascinating oil beetle, whose larvae are parasitic on solitary bees, is occasional, along with the brilliant green rose chafers. Graylings are perhaps the most coastal of our butterflies, but others include green hairstreak, dark green fritillary, brown argus and small blue. The coastal scrub also provides ideal cover for the badger, and many setts are found on the cliffs of West Wales.

The soft coasts of Carmarthen Bay and the Dyfi estuary are noted for their specialised sand dune and salt marsh plants, while for most of the year offshore shallow waters provide feeding and roosting grounds for important flocks of common scoter. In March 1974, an estimated 25,000 were observed in Carmarthen Bay. They dive to feed, in depths of 6m or sometimes even 9m, on the rich molluscan infauna that marine scientists have observed in the fine muddy sand of the Bay. As early as 1932, Bertram Lloyd recorded the presence of big summer flocks, after voyaging across Carmarthen Bay in a fishing smack. The queries he noted in his journals are still being answered: 'Are these birds here *all* breeders (very likely). Splendid shallows and shoals here for their feeding purposes (diving)'. The wide sandy foreshores of Carmarthen Bay have long been a favoured haunt of sea-shell collectors. Sandy beaches extend westward as far as Saundersfoot and Tenby. The beer-barrel shell is particularly common at Pendine, while dense populations of razor-shells, together with the pod-razor are found south of Saundersfoot, where they have been subjected to relentless commercial exploitation. Cefn Sidan sands and Pendine sands are favoured by oystercatchers and other waders, with over 5,000 roosting at Towyn Point during autumn migration.

Broken sea-shells are an important component of the marine coastal dunes that border Carmarthen Bay, and are largely responsible for the high lime content of most of the dunes of West Wales. Large, lime-rich dunes dominated by marram grass, like those at Ynyslas, Pendine and Pembrey, are a paradise for plant lovers from spring through the summer, when scented purple carpets of wild thyme vie with the yellows of lesser hawkbits and the pink of common restharrow. Bee orchids, pyramidal orchids and autumn lady's-tresses are other delights. The menacing henbane occurs sporadically after disturbance of surface soil, along with the yellow horned-poppy. The last named plant is more usually associated with shingle vegetation but, apart from a few traces at Ynyslas and the Gann Estuary, there is little in West Wales.

The hollows with their high water tables are known as dune slacks. They are particularly suitable for orchids, and the early, northern and southern marsh-orchids and the marsh helleborine often grow in great profusion in the larger slacks, provided woody growth has been suppressed by rabbits. Other notable plants include two Red Data Book plants — the fen orchid now confined to Pendine dunes, and the more widespread dune gentian. Pendine, where access is strictly controlled by the Ministry of Defence, is the most outstanding dune system, with its extensive dune slacks and wide transition to fen around the shallow Witchett Pool. These dunes around Carmarthen Bay abound in butterflies, especially the marbled white, and are noted for the abundance of the rare beetle *Eurynebria complanata* which breeds among the debris at the seaward toe of the dunes. Unfortunately, swathes of sea buckthorn are progressively spreading over large areas of these dunes, impoverishing their flora and fauna. A similar fate has befallen the southern portion of Tenby dunes. Extreme exposure and management as rabbit warrens, prior to the myxomatosis epidemic of the late 1950s, has kept the dunes of the Castlemartin peninsula free of scrub. They reach inland over the cliffs, forming extensive dune pastures rich in rare lichens, with those at Stackpole graded as of international importance.

In the sheltered inlets of the Dyfi, Milford Haven and the three rivers of Carmarthen Bay, extensive mud flats are exposed at each low tide. Flowering plants can colonise even the lowest flats. Narrow-leaved eel grass covers some 100 hectares of the lower level flats in Milford Haven, principally in the embayments of Angle and Pembroke river. Its larger relative, eelgrass, grows on mud submerged by two metres of water at low tide between Littlewick and Gelliswick Bays. The main channel of Milford Haven is of special interest to the marine ecologist owing to its depth and the proportion of hard substrate, as well as the effect of freshwater inflows. The middle reaches above the ferry bridge experience a wide salinity range, while the upper reaches above Llangwn have low salinity levels. Work by Swansea University on the sublittoral fauna has shown a loss of a quarter by the middle reaches and a loss of a half of the species once the upper reaches are entered, compared with the fully saline areas below the bridge. A similar decrease in diversity occurs on the shores. Away from the peculiarities of the main channel of Milford Haven, its arms and embayments, like those of other West Wales estuaries, have their upper flats colonised by salt marsh. Over half is dominated by common cord-grass which has spread rapidly in the last 40 years, as it has elsewhere in Britain. By 1983, however, numerous sections of Milford Haven and the Carmarthen Bay estuaries were showing signs of cord-grass dieback, perhaps due to water-logged conditions. Of the 390 hectares of salt marsh in Pembrokeshire, 200 hectares were cord-grass in 1983. Only the diminutive Newport estuary remains free of the grass, and so far its sporadic attempts to colonise have been repelled by judicious use of bucket and spade!

More diverse salt marsh occurs at higher levels, especially where the vegetation is ungrazed and where there is sand and shingle as well as mud. In the most mature saltings, the variety is further enhanced by salt marsh pans and deep creeks, as on the Dyfi, the Gann and the Gwendraeth, Tywi and Taf. Creeks are often lined with sea-purslane shrubs, while nectar-seeking insects gorge themselves on the adjacent flowers of sea lavender and thrift. Common sea-lavender grows on the Carmarthen saltings, while lax-flowered sea-lavender is found in Pembrokeshire, and in both areas, on sandy or stony margins, stands of the rock sea-lavender occur. It is in these upper reaches of the saltings that the plant hunter may be rewarded with the marsh-mallow, with sharp rush which forms two metre high clumps that more than live up to their name, or for the real enthusiast, scarce glassworts like *Salicornia pusilla*. The Newport saltings are the southernmost location in Britain for the saltmarsh flat-sedge. A recent discovery of a more inspiring plant was the finding of the rare marsh peas on the saltings of the Eastern Cleddau in 1982. This attractive plant has only one other location in the rest of Wales at the Ffrwd Farm Nature Reserve near Pembrey. In such upper reaches the salt marsh is highly influenced by freshwater. Beds of common reed are frequent and, in the least modified places, especially on the Dyfi, the aromatic bog myrtle creeps into the edge of the saltings. To many, the abiding impression of mud flats and saltings lies not just in the play of light or the addictive smells, but in the haunting cries of the waders. Prolific food supplies are offered by muddy shores and these are exploited by the large flocks of waders and wildfowl that rest while on passage, or winter on the relatively warm coast of West Wales.

Dunlin, curlew and redshank are the main estuarine waders, while shelduck, widgeon, mallard and teal are the most numerous wildfowl. The Dyfi is perhaps the best for widgeon, and the Carmarthen Bay estuaries for dunlin and also for oystercatchers. More concerted counts of the Taf Tywi and Gwendraeth are required if their true ornithological signifiance is to be appreciated. Between 1982 and 1986 monthly counts have been undertaken on the same day throughout the whole of Milford Haven, by a team of as many as 15 enthusiasts. This has shown that numbers of teal, shelduck and curlew have exceeded national significance on a total of 24 occasions, while redshank and widgeon have been at national levels twice. Mid-winter totals for duck in the Haven regularly surpass 5,000, and for waders they fluctuate between 6,000 and 10,000. Only through the dedicated teamwork of these Trust members has the importance of Milford Haven been fully appreciated. It is such voluntary efforts that have added so much to our knowledge of the coast of West Wales.

Grey Seal Pup

ABOVE: Strumble Head in May. (SBE) LEFT: Marsh-mallow on shore of Milford Haven. (JWD) CENTRE: Rock samphire on east cliffs of Ramsey. (SBE) RIGHT: Rock sea-spurrey on Ceredigion coast. (JH)

PLATE V

ABOVE: Skomer from 1700m. CENTRE LEFT: Kittiwake and young at nest on Skomer. RIGHT: Razorbills. BELOW LEFT: Grey seal pup several days old. RIGHT: Oystercatcher. (All MA)

PLATE VI

ABOVE: Ceibwr Bay, north Pembrokeshire. (SL) BELOW: Castlemartin Peninsula, the cliffs eastwards towards Bulliber Down. (MOD)

ABOVE: Milford Haven waterway from the Cleddau Bridge; the estuaries and creeks adjacent to the main waterway are nationally important for some waders and wildfowl. (SL) BELOW: Saltings on the Dyfi estuary. (JPS)

ABOVE: Dune slack at Ynyslas. (JPS) BELOW: Removing encroaching common cord grass on the Nevern estuary. (SBE)

ABOVE: Board walk at Poppit Sands to reduce erosion. (SBE) BELOW: Sand extraction at Broomhill Burrows. (SBE)

ABOVE: Juniper on the south cliffs of Ramsey Island. (SBE) BELOW LEFT: Roseroot, an arctic-alpine found on St David's Head. (NCC) RIGHT: Spring squill, which carpets some cliff tops in May. (JPS)

LEFT: Sea rocket, found on many sandy beaches. (NCC) RIGHT: Pioneer naturalist, the late T.A.W. Davis, standing beside sharp rush which occurs on the saltings and dune slacks of Carmarthenshire and Pembrokeshire. (SBE) BELOW: Grey seal haul out on a Pembrokeshire beach. (SBE)

ABOVE: Oystercatchers, one of the most numerous waders on the Burry Inlet. (HG) CENTRE: Peregrine, now returning to many former haunts. (HP) BELOW: Choughs, virtually restricted to the coast in West Wales, save for several pairs inland in North Ceredigion. (HP)

ABOVE: Skokholm from the north-east. (WWTNC) BELOW: St Margaret's with Pembrokeshire mainland in background as seen from Caldey. (JF)

Islands

Wheatear on Skokholm

From the mists of the Dark Ages, when Scandinavian sea-warriors passed this way, leaving as evidence the place names, the islands off the south-west coast of Wales emerge as places of agricultural activity, where incomes could be augmented by a crop of the newly introduced rabbit and even seabirds. The farmhouses and field systems were largely developed during the 18th century, and from then we have an increasingly clear picture of the islands and their inhabitants. At the end of the 19th century, information on island natural history becomes available and their importance is recognised. Now the nature conservation aspect is paramount on five of the islands, and the sixth, Ramsey, although some farming does continue, has been designated a Site of Special Scientific Importance. Let each island speak for itself.

Caldey, together with its satellite St Margaret's, is situated about 3.5 kilometres south of Tenby and about 1 kilometre from Giltar Point, the nearest point of the Pembrokeshire mainland. Although St Margaret's and its birds present a microcosm of the larger western islands of Ramsey, Skokolm and Skomer, Caldey is different, despite being the third largest, (some 255 hectares in extent).

The island is comprised in the north of Carboniferous Limestone and the south by Old Red Sandstone and has a coastline more varied than the other islands. In addition to rocky shores, cliffs and sidings, Caldey boasts several fine sandy beaches backed in places by dunes. The interior slopes gently northwards from the highest points of about 60m, close to the southern cliffs, thus affording a reasonable degree of shelter from prevailing winds. Much of the interior is farmed, and the only wild areas are around the cliff tops outside the field boundaries and on the scrub-covered north-eastern headland. Also quite unlike the other islands, are the coniferous plantation, stands of deciduous trees and fuschia hedges, while a farm, hamlet and Cistercian monastery are other features which make Caldey unique.

The caves in the Caldey cliffs have provided many clues to the presence of early man and of some of the animals which roamed the plains and forests of the drowned lands beneath the Bristol Channel. About 1840, quarrymen broke through into a cave near Eel Point, and found a quantity of bones most of which they sent to Bideford for sale as fertiliser. Fortunately Richard Greaves, a visitor to Caldey, realised their significance and informed the Geological Society of London. Subsequent searches in the caves have revealed the remains of a rich mammal fauna, including mammoth, rhinoceros, bison, reindeer, lion, hyaena and the aurochs.

The bird life of Caldey is quite different compared to the other Pembrokeshire islands, with fewer seabirds, but the range of inland habitats means a much greater range of breeding land birds. These include shelduck, collared dove, swift, several species of tit, tree creeper, mistle thrush, blackcap, greenfinch and yellowhammer.

Seabird population (in pairs) on main Pembrokeshire islands

	Caldey	Ramsey	Skokholm	Skomer
Fulmar	24	102	18	472
Manx Shearwater	-	?	35,000	100,000
Storm Petrel	-	-	6,000	1,000
Cormorant	-	-	-	20
Shag	7	10	-	3
Lesser Black-backed Gull	90	120	4,200	13,200
Herring Gull	680	-	350	725
Great Black-backed Gull	1	?	26	11
Kittiwake	-	350	-	2,380
Guillemot	25	1,600	160	6,180
Razorbill	22	850	400	3,580
Puffin	1?	-	2,500	7,000

Herring gulls, the most numerous seabird on Caldey, perhaps the result of rich feeding during summer on the nearby holiday beaches, have declined dramatically in recent years. In 1970 there were some 3,250 pairs, in 1975 3,860, but since then numbers have dropped and in 1985 only 684 were nesting. Similar reductions have taken place at the other Pembrokeshire islands — part of a general decline in the herring gull through much of Great Britain. The cause may be botulism, contracted through feeding on rubbish dumps, where household rubbish in black plastic bags forms in warm weather an ideal incubation situation for the bacillus.

Most surprising is the paucity of small mammals, in view of the high degree of human occupation, affording greater opportunities for accidental introductions. There were rabbits in the past, but were they there by the early 14th century? The earliest reference traced by that assiduous historian of the Pembrokeshire islands, Roscoe Howells, is 1579. Rabbits were exterminated on Caldey in 1975 by much human endeavour, following a major outbreak of myxomatosis a year or two before. The only other land mammal present is the brown rat, which has been present for an unknown period, possibly the result of a shipwreck in the last century.

Cardigan Island, the only island in Dyfed outside Pembrokeshire, is on the north side of the mouth of the Teifi. Some 15 hectares in extent, a narrow channel some 150m wide separates the island from the mainland — close enough to encourage one promoter in 1907 to consider erecting a cable way to provide a means of transit to the island. There is a good coastal footpath commencing beside the Cliff Hotel at Gwbert. Rather little is known about the history of the island, though there is evidence of early human activity. A low turf wall encloses the site of what may be the remains of an early Christian cell, while there is a dew pond cut deep into the rock at the island centre. In the 13th century the island was part of the King's lands at Cardigan, and known as Hastiholm. Sheep were traditionally grazed there, a practice which continued until the 1930s.

To perpetuate sheep grazing and at the same time help in the conservation of a then rare breed, the Trust introduced Soay sheep from the Duke of Bedford's flock at Woburn, then one of the few flocks in existence other than the original one on St Kilda. The descendants of this original introduction still roam the island, and on occasions number just over 100 animals. Their continual presence has significantly moulded the vegetation, indeed some experts feel it has been impoverished and as a result the flock should be removed. This is open to much debate and meanwhile a programme of scientific work has commenced. In addition a catching pen has been erected with the aid of a Nature Conservancy Council grant. This will enable the sheep to be regularly caught and ear tagged, so that more can be learnt of their population dynamics.

The bird population is dominated by the large gulls. Herring gulls are the most numerous species, with up to 1,000 pairs, but in 1984 and 1985 they declined and now number some 500-600 pairs. A small number of great black-backed gulls nest, in recent years no more than 20 pairs. Lesser black-backed gulls probably only commenced nesting in the 1960s and now number about 300 pairs, occupying much of the central plateau. Razorbills began to land on the island in 1982 and have bred since 1983. Other birds breeding include fulmar, shag, oystercatcher, raven and rock pipit, while skylark and meadow pipit used to, but now seem to be excluded by the nesting gulls.

Ten years before it became a nature reserve the very worst calamity which can befall a seabird island took place. On a stormy night in 1934 the liner *Herefordshire* of the Bibby Line broke her tow while being taken to Glasgow for breaking up. At about 07.00 she drifted on to Cardigan Island close to the north west point, the towage crew of four eventually scrambling to safety onto the island, rescued later in the morning by breeches buoy. Also scrambling ashore were brown rats, ruthless predators on the eggs and chicks of burrow nesting seabirds.

That a puffin colony once existed on Cardigan Island is not in doubt, but information as to its extent is scanty. It is known to have existed in the 1890s, the birds described in a local guide book as 'Welsh parrots'. A sheep grazier early in the present century referred to the island as full of puffins, but the only firm details seem to be that in 1924 the colony was between 25 and 30 pairs strong. They eventually died out, their demise undoubtedly hastened if not caused by the depredations of the rats, the population of which at one time was estimated at 1,200 animals.

In the hope of encouraging puffins to return, the Trust, aided by pest control officers from the Ministry of Agriculture, managed to clear the island of rats in 1968-1969. No puffins or Manx shearwaters took up the offer, so after ten years the Trust initiated further activity.

First burrows were dug, mainly by a party of scouts under the direction of Aberystwyth naturalist Dr A.D.Q. Agnew. Then, commencing in the late summer of 1980, batches of 50 fledgling Manx shearwaters, by then independent of their parents, were taken annually from the enormous colony on Skomer, transported rapidly to Cardigan Island and placed in the burrows. The hope was that, when the fledglings emerged, in the course of the next few nights, they would imprint Cardigan Island into their navigation systems and in due course return there, rather than their natal home on Skomer. In 1984 work was rewarded by the discovery of a Manx shearwater egg in one burrow, though unfortunately it was deserted. In 1985 this and other burrows showed signs of visitation or were freshly excavated by the birds and one hopes that it will not be long before a colony becomes established.

What of the puffin, the original participant in the drama of the *Herefordshire* and its stowaway rats? Their breeding cycle does not permit the transfer of young, so that another method is being tried. In 1984 a large number of wooden decoy puffins were placed on the cliff slopes close to the burrows of the potential colony. It is hoped that birds flighting or resting on the sea offshore, as they sometimes do in spring, will be lured to investigate, find empty burrows and so re-establish themselves. However, the large gull populations, now occupying much of the surface of the island, may pose something of a problem for future colonists.

Some 10 kilometres due west of Skomer is the tiny island of **Grassholm,** just 9 hectares in extent but easily visible in clear weather from the great sweep of St Bride's Bay, or southwards along the red cliffs of Dale and across lonely Linney Head. Grassholm is unique in Wales, for it contains the only gannetry in the Principality.

Like much of the island's early history, that of the gannetry is shrouded in mystery. Certainly gannets were there in small numbers in the 1860s and, according to the old fishermen of Marloes who could remember such things, were there perhaps as early as the 1820s. Had gannets nested previously? Surely the early chroniclers of Pembrokeshire would have made some comment?

In the beginning the gannets were regularly raided for their eggs and in at least one year adults were killed for bait. On another occasion a party of army officers shot adult gannets, killed young on the nests and smashed eggs. Fortunately, the perpetrators were brought to trial, found guilty and fined, one of the first cases under the Wild Birds Protection Act of 1880. Despite the hazards, the colony was increasing, when the first historian of the gannet, the Norfolk ornithologist J.H. Gurney sailed by in 1903, for it was too rough to land; there were between 250 and 300 pairs. Since then, and despite fire, use as a bombing target in World War II, oil pollution and birds becoming trapped in discarded fishing line and nets, numbers have increased with hardly a check.

Numbers of Gannets (in pairs) breeding on Grassholm

1820	?	1903	275	1949	9200
1860	20	1905	275	1951	8000
1883	20	1907	300	1952	8500
1886	250	1922	900	1956	10550
1889	200	1924	1900	1964	15500
1893	240	1933	4750	1969	16128
1895	300	1939	5875	1875	18350
				1984	28000

Colonel H. Morrey Salmon, father-figure of Welsh ornithology, together with Miss C.M. Acland, made a pioneering photographic survey of the colony in 1924. Like all visits to Grassholm their visit was much influenced by the weather, though few visitors will have been at sea for such a period. They left the mainland at 1.30 a.m., were forced back by freshening weather which then moderated, made a second start at 10.30 a.m. and returned 12 hours later, all for a bare half hour on the island.

More comfortable but equally pioneering was the first aerial census. In September 1956 R.M. Lockley and J. Radford flew low over the island in an Auster aircraft taking a series of photographs with a large mapping camera. Subsequent surveys have all been from the air; the most recent in 1984 by M. Alexander revealed a population of 28,000 pairs. Grassholm is now the second largest gannetry in the North Atlantic after St Kilda. The rate of increase has been such that immigrants from other colonies, most probably those on Little Skellig off Co Kerry and Ailsa Craig, sentinel of the Clyde, have contributed to the success. Now Grassholm and its gannet colony is one of the greatest wildlife spectacles of north-west Europe. No one can fail to be moved on seeing the vast concourse of birds and listening to their harsh clamour.

During the last century Grassholm was home for an enormous number of puffins, though their subsequent fortunes have been the reverse of the gannetry. Some estimates put the population at half a million birds and, although there would have hardly been room for such a gathering, it may well have totalled 100,000 pairs. Such numbers dominated the island as 'We were hard put to find a bit of ground clear of puffins-burrows on which to erect our very small tent, but having as we thought, done so, we put the tent up. At about 3 a.m., I was awakened by a curious grunting sound under my ear, and on removing my pillow, there was Mr or Mrs Puffin, with a very grieved expression, sitting at the entrance of a burrow. I made the *amende honorable* by clearing a way to the entrance of the tent, down which the aggrieved party attempted to make a dignified exit. I mention the above to give you an idea how thick the puffins were on the ground in these days'.

A rapid but sadly undocumented decline took place during the early part of the present century for, in 1934, only 130 pairs were present. By 1948 the number had almost halved, and little more than ten years later only two or three pairs remained, a situation which continues to the present day. Most observers attribute the demise of the puffin colony on Grassholm to the collapse of nesting burrows, a labyrinth of tunnels in the thick peat of decaying red fescue, the dominant plant over most of this ungrazed island. Nowadays, the burrow system, in the words of one visitor 'a puffin Pompeii', is still plainly visible close to the edge of the expanding gannetry. Elsewhere the red fescue carpets the roofless burrows, making walking difficult for the unwary. As the burrows became untenable, so the puffins moved to Skokolm and Skomer where the colonies rapidly expanded during the first decades of the century.

Alas, the roseate terns which nested on Grassholm a hundred years or so ago have long since vanished, but other species remain. On the north-facing cliffs, which they share with the gannets, are small numbers of kittiwakes and guillemots, while elsewhere there are razorbills and, hiding among some boulders, a few shags. All three species of large gulls are present, though the populations of both the great black-backed and herring gull have declined in recent years. The only other breeding birds are the oystercatcher, rock pipit and a pair of ravens. Do the latter find enough food on the island for their chicks, or do they need to visit the avian supermarkets of the other islands or the mainland? Graham Williams, ornihistorian of Grassholm writing in 1978, recorded in addition to breeding birds no less than 78 casual visitors. None of the occasions on which I have been to Grassholm has been without such treasures to be found on seeking a quiet spot away from the gannets. I well remember the red-breasted flycatcher hawking for midges on the damp cliff above the south-west landing. That great October day when there was a pintail on a rock pool, a short-eared owl was flushed and other birds present were wood pigeon, rook, black redstart and snow bunting.

Stand on the crest of Grassholm and look through the whirling gannets which circle in a never-ending stream and you will see The Smalls, a low reef some 13 kilometres to the west. This is the uttermost part of Wales, indeed it is further west than the coasts of Co Down. A lighthouse has stood guard on The Smalls since 1776 but that is another story. The rocks are completely awash in rough weather so that no birds breed there. Lighthouse keepers have from time to time kept records of passing birds, and those which seek brief shelter around the lantern rails. The main feature of nature conservation interest however is the large number of grey seals from the breeding colonies on the larger islands and mainland coast which lie up here.

Between The Smalls and Grassholm are reefs known as the Hats and Barrels. The Hats, some 3.5 kilometres east of The Smalls, remain completely covered; for those that venture close on the few occasions when sea conditions permit, only the tips of the *Laminaria* fronds brush the surface even at low spring tides. The Barrels, 3.5 kilometres further east, dry out by 3m at low water. It was here that both the ill-fated *Christos Bitas* in October 1978 and the *Bridgeness* in June 1985 ran aground, the resultant oil pollution causing the deaths of many thousands of seabirds and, in the case of the *Christos Bitas,* also some young seals.

Less is known about the natural history of **Ramsey** than the other western islands of Pembrokeshire. It merits a mere five pages in *A Natural History Bibliography of Pembrokeshire* published in 1980, the same as tiny Grassholm, compared to 30 pages for Skomer and no less than 42 for Skokholm. This may be partly explained by the fact that, except for the period 1964 to 1976 when the island was managed by the Royal Society for the Protection of Birds, visits by naturalists, or at least naturalists who have subsequently put pen to paper, have been few.

Scientifically Ramsey is the most dramatic of the islands, as a boat journey around it will confirm. Some three kilometres north to south and extending for about 264 hectares, the island comprises a gently sloping plateau on the east, which quickly gives way to steep heather-clad rocky hills on the west. Carn Llundain in the south rises to 136m and is divided from the slightly lower Carn Ysgubor in the north by the wide expanse of Aber Mawr. Below the hills are dramatic cliffs and chasms.

Beyond the island are rocks and stacks, Ynys Berry, guarding the southern flank, an island in its own right. Between each runs fast wind riot tide races, that of Twll-y-dillyn between Ynys Berry and Ynys-y-Cantwr being not much wider than the boat, or so it seems as you hurtle through, the cliffs towering silently above. To the west are the Bishops and Clerks, four main inlets and numerous reefs which, in the words of the Elizabethan historian George Owen 'preach deadly doctrine to their winter audience and are commendable in nothing but for their good refidence, thefe all yelde ftore of gulls in the tyme of the year'. Small wonder that a lighthouse was built on the South Bishop rock in 1839 to warn shipping against approaching this dangerous section of coast, and the St Davids lifeboat has carried out many a brave service around Ramsey and these rocks.

The written history of Ramsey extends back to the 14th century, that of legend into the mists of the Dark Ages, almost to the beginning of Christendom. In the late 2nd century AD, Devanus preached throughout the land and retired to end his days on Ramsey. Several centuries later, Justinian, confessor to the saintly David, had his chapel on the island and was eventually murdered there by some of his mutinous followers. They were then struck with leprosy and sent to work their penance on Ynys y Gwahan, the Lepers Island to the north of Ramsey Sound. One of the delightful legends of St Justinian concerns how he came to settle on Ramsey. Obviously troubled by his followers, he withdrew to the island along a narrow isthmus which still connected it to the mainland. With a mighty axe he cut away at the rocks so that the sea began to pour through to provide a barrier against the problems of the mainland. Unfortunately his axe, (or was it that his strength was not quite equal to the task), became quite blunt and curtailed his labours so that a line of rocks, the Bitches, still extends eastwards into Ramsey Sound.

Ramsey emerges from the Dark Ages when an inventory made in 1293 of the Bishop of St Davids' goods provides the first fascinating details: 'There are in the island 5 steers worth 15s at 3s per head. 8 heifers worth 20s at 2s 6d each. 2 bulls of 2-years old worth 4s. 9 bullocks and young heifers worth 13s 6d at 18d each. 200 one-year old [calves] worth 100s at 6d per head. 44 muttons with wool worth 36s 8d at 10d each. 70 goats worth 35s at 6d each. Total £6 13 2.

'There are there, 10 acres of wheat sown worth 25s at 2s 6d per acre. 3 acres and 1 stang [rood] of barley worth 9s 9d at 3s per acre. 17 acres of oats worth 2s 4d per acre. 12 cribs of oat malt worth 48s at 4s per crib. Total 111s 1d.'

Nearly 30 years later *The Black Book of St. Davids* gives further information, which includes reference to 100 loads of rushes and heath at 3d a load taken from the island, and also the first mention of rabbits, some 500 valued at 33s 4d. After this several centuries were to elapse with little written documentation, though without doubt farming and the taking of rabbits would have continued. Indeed farming survives right to the present day, with sheep the main agricultural activity.

In 1979 red deer were introduced in an effort to improve the island economy, as a deer farm. The rabbit, a major source of income from medieval times up to the early 1950s, is now seen as a pest, with much effort made over the past three decades to eradicate or at least contain it. The struggle continues.

A mammal which has played its part in shaping the island's avifauna is the brown rat, which reached Great Britain in the early 18th century. It seems likely that the rat became established on Ramsey about 1800, though how it arrived is a mystery. The brown rat would certainly have been responsible for the demise of the puffin colony, which in 1715 was described as vast. They were taken for food at this time by the citizens of St Davids, who ate them during Lent on account of their fishy taste. *The Universal British Traveller* published in 1776 also contains a reference to puffins when it reports that together with harrybirds (Manx shearwaters), it 'bred in holes and commonly those of the rabbit'. Shortly after this the rat arrived, for Richard Fenton, writing in 1811, records that even the rabbit population had been reduced by the rats. By the late 19th century the only puffins were confined to the north end of Ramsey, with a few pairs into the 1900s at inaccessible points on the cliffs; now there are none. Nothing is known of the fortunes of the Manx shearwater, their current status a mystery. Does the occasional pair still nest? The cliff nesting seabirds — kittiwakes, razorbills and guillemots — are mainly confined to the west coast. Compared to the other islands, the number of gulls nesting is miniscule and these are largely confined to Ynys Berry off the south end of the island. Fulmars, shags and cormorants also nest.

An important bird on Ramsey is the chough, which nests in cliff crevices or on cave ledges. In 1972 the population was eight pairs together with two non-breeding pairs, a fifth of the Pembrokeshire population. A study by Mrs Susan Cowdy in June and July that year showed the main food was ants and their larvae, mainly obtained by the birds probing in the short close-cropped turf. This terrain is essential to the chough, and a drastic reduction in sheep levels on Ramsey during the past fifteen years has allowed taller vegetation, particularly heather and bracken, to become dominant; the result, a halving of the chough population., Quite the reverse has occurred on Bardsey in Gwynedd, and on the Calf of Man, where increased grazing has resulted in increases in chough numbers. Among other birds, the lapwing in 1985 had no less than 36 pairs nesting in the grazing fields on the east of the island.

Ramsey is the largest seal nursery in southern Britain. About 300 pups are born each autumn, mainly between early September and the end of November, with the peak in October. Open beaches like those of Aber Mawr, Aber Garlic and Porth Llauog are used, as well as other beaches deep inside sea caves like Ogof Tywod, Ogof Velvet and Ogof Dafydd. Seals are present around Ramsey throughout the year, with numbers increasing in August, when they begin to take up station at the breeding beaches, the bulls arriving first, followed by the cows. Much pioneering study was carried out on Ramsey into the breeding of the grey seal, at first by J.L. Davies, followed by H.R. Hewer, R.M. Lockley and others. An early attempt at marking seals was carried out on Ramsey in 1946 using a branding technique. Later, various types of tags were used, and as a result a good deal was learned of the dispersal of pups after they leave the natal beach. There were records north into the Irish Sea, around the south and west of Ireland to Co Galway, to the south-west of England and Brittany, while one remarkable traveller reached the north coast of Spain.

Connected to the northwest corner of Caldey Island by a reef uncovered only at low tide is **St Margaret's Island.** In the 16th century, naturalist John Ray referred to the island having a small chapel consecrated to St Margaret, and it has been subsequently known by that name, though previously it had been known as Little Caldey.

The island is mainly a narrow ridge of Carboniferous rock rising some 35m at sea level, and some 7 hectares in extent. Towards the eastern end is a quarry which nearly intersects the island; operations at the quarry ceased about the middle of the last century, although the remains of several buildings used by quarry workers and their families survive. The stone was sent over a wide area of south-west Wales, its main use being as burnt lime for agricultural use.

The island has been leased to the Trust since 1950, the main feature of interest being the largest cormorant colony in Wales, itself among the largest in Great Britain. About 200 pairs have nested regularly since 1969, with a peak of 322 pairs in 1973. Over 3,000 young cormorants have been ringed under the direction of S.J. Sutcliffe, who has made numerous visits annually since his first as a schoolboy in 1962. Ringing recoveries from this ambitious project now number several hundred, with recoveries from as far afield as Spain and Portugal, although most are from the south coast of England and the Bay of Biscay, together with many in local waters. Wynford Vaughan Thomas may well have had the easily viewed colony on St Margaret's in mind when he wrote that 'The cormorants are my favourite sea bird. They clustered on the edge of the rocks black suited and looking like a group of deacons discussing the sermon after chapel. Or suddenly changing character as they spread their wings out to dry — miniature Count Draculas'.

Despite its small size, a number of observers have visited the island, including John Ray, who crossed the island in 1662, noting that the tree mallow was abundant and that among birds the nests of the 'puits and gulls and sea swallows lying so thick that a man can scarce walk but he must need set his foot upon them'. Alas, the sea swallows [terns] have not nested here in modern times.

Much more recently John Walpole-Bond describes a visit in 1902 in that almost forgotten classic *Bird Life in Wild Wales* (1903), and refers to puffins in such a manner as to suggest a colony of some size. A photograph in the book shows some 29 birds on a small section of cliff top, and one wonders just how many birds were elsewhere on the island. Now only about three pairs nest in cliff crevices.

Bertram Lloyd, the assiduous diarist, landed on 6 June 1927 having been rowed from Caldey,'in a leaky Canadian canoe only capable of holding two persons with difficulty'. The main object of his visit was to search for Manx shearwaters, though none was found during 'a horribly wet night' before 'we crossed the gut to Caldey again in the comic little canoe' the following morning.

Gulls nesting on St Margaret's Island

	Great Black-backed Gull	Lesser Black-backed Gull	Herring Gull
1970	150	11	600
1971	124	16	535
1972	126	12	592
1973	134	12	545
1974	127	7	653
1975	128	7	628
1976	147	7	513
1977	128	6	564
1978	109	10	478
1979	148	10	402
1980	114	11	388
1981	n.c.	12	205
1982	71	10	110
1983	n.c.	11	-
1984	60	14	137
1985	44	11	108

On the cliffs there are colonies of kittiwakes, razorbills and guillemots, while the plateau is occupied by all three large gulls. In 1985 there were but 44 pairs of great black-backed gulls and 108 pairs of herring gulls. This compares with much higher numbers in the early 1970s.

As with Skomer, the discovery of flint-flakes takes the history of **Skokholm** back to a time in the Middle Stone Age, perhaps even earlier, when a few hunter-fisherfolk passed this way. A subsequent rise in sea-level following the last Ice-Age was sufficient to isolate the island, and to prevent its colonisation by Iron Age farmers as occurred on Skomer. From medieval times until the early 20th century both islands followed much the same pattern of pastoral activity and rabbit catching. Indeed, for long periods the islands were in the same ownership or tenancy. Since the mid-18th century Skokholm has been part of the Dale Castle Estate. This continuity of ownership is to be applauded.

In 1927 a lease of Skokholm was granted for 21 years to Ronald M. Lockley who, since boyhood, had dreamed of living on an island and of making a study of its bird life. This event was the foundation on which much that happened subsequently on Skokholm, and indeed the other islands, was built. Not only did Ronald M. Lockley make a number of pioneering studies of seabird life history, but in 1933 he established the first British Bird Observatory. More than all this Ronald M. Lockley has been the author of many natural history books, but his first, those concerning Skokholm, like *Way to an Island, Dream Island* and *Letters from Skokholm,* have fired countless naturalists with an enthusiasm for island-going.

In 1940 the Lockley family moved from Skokholm as a result of the war, and the island remained largely unvisited until 1946, when the Bird Observatory re-opened. A succession of wardens was employed, and these continued, with the seabird studies and the ringing of bird migrants. It was a major disappointment when in 1976 the Observatory was forced to close and ringing ceased, though the island is still managed as a nature reserve with a warden, and with opportunities for visitors to enjoy a holiday. From 1947 to 1969 the island was administered from Dale Fort Field Centre and for all but the last year was the responsibility of the Dale Fort Warden, John Barrett, who so strikingly sums up his 21 years in the foreword to the *Observatory Report* for 1968 when he writes: 'Having started in June 1947, my responsibility for the maintenance, supply and administration of Skokholm ended on October 19th 1968, when David Emerson took over from me as Warden of Dale

Six examples of West Wales butterflies: ABOVE LEFT: The large heath at its most southerly location, Cors Caron in Ceredigion. RIGHT: Small copper, a widely distributed species. CENTRE LEFT: Marsh fritillary, a declining species nationally, but still locally abundant in West Wales. RIGHT: The brown hairstreak is a local butterfly in West Wales. BELOW LEFT: The grayling occurs mainly in coastal districts. RIGHT: The brimstone breeds locally in Carmarthenshire and Ceredigion but not in Pembrokeshire. (All JPS)

PLATE VII

Eight examples of damsel and dragonflies found in West Wales. ABOVE LEFT: Male Common darter, a widely distributed later summer species. RIGHT: Male broad-bodied chaser which may be seen at most large ponds from mid-May onwards UPPER CENTRE LEFT: Beautiful demoiselle, a jewel of a damselfly, found beside many fast flowing streams. RIGHT: Small red-damselfly, a local species of bogs and heathy ponds. LOWER CENTRE LEFT: Male keeled skimmer, breeds in *Sphagnum* bogs. RIGHT: Variable damselfly, a scarce species in West Wales. BELOW LEFT: Golden-ringed dragonfly, found throughout West Wales. RIGHT: Black darter, usually found on bogs and heather areas. (All JPS)

Fort. In those twenty-one years two or three thousand people have been taken to the island and I wonder how many tons of stores and luggage. I have once seen a "freak" wave (very frightening), once a school of killer whales (very powerful) and once a shearwater so rare that nobody would believe me (very properly). In good times the boat has left at 8 a.m. fourteen consecutive Saturdays, in bad times we have had ten travellers stranded in the Fort which was already overfull of its own students. Twelve days overdue was the longest hold-up.'

Skokholm is a flat-topped island of Old Red Sandstone lying three kilometres south of Skomer and three kilometres west of the mainland cliffs of the Marloes peninsula. Some 100 hectares in area, the island slopes gently from an altitude of 50m at the western end to the 20m of the cliff-tops in the east. There is little relief on the top of the island, except for a few ridges of rock which rise about 10m above the surrounding plateau. Most of the circumference is precipitous, but access to the shoreline is possible in several places and two coves provide sheltered landings; South Haven has a small jetty and an access road, and is used for landing the majority of visitors and supplies.

Gannets

For the most part, the soils of Skokholm are light sandy loams, but a band of boulder clay traverses the centre of the island and the impaired drainage gives rise to a large boggy area and two semi-permanent ponds. Two small bogs, East Bog and Orchid Bog, lie on a spring-line, as does a shallow but reliable well near the buildings.

A history of cultivation, land-drainage and grazing by introduced mammals has had a profound effect on the vegetation. Currently there are almost no woody plants over 25cm high, but a complex mosaic of herbaceous stands, which form seven major communities, and a number of minor associations. The diversity of the vegetation is in part maintained by environmental factors, such as exposure and water availability, but these communities whose distribution is largely controlled by factors such as grazing and trampling are particularly prone to fluctuation and mutual exclusion. Trampling, grazing by rabbits, and burrowing by puffins, shearwaters and rabbits can lead to destruction of vegetation and topsoil and often result in local erosion.

The more sheltered cliffs provide a haven for succulent halophytes like sea plantain, rock samphire and tree mallow, that would otherwise be eliminated by rabbit grazing. On the higher slopes, and extending inland by varying degrees depending on exposure, is a band of maritime heath from Crab Bay west to the lighthouse and then along the north-west coast to Little Bay. This habitat is dominated by thrift, sea campion and scentless mayweed, and is one of the sights of the summer. In good sea campion years, Skokholm from a distance seems ringed with a halo of white. Red fescue is also a common component of the maritime heath but on Skokholm it is heavily grazed: in the absence of rabbits, as on Grassholm, it becomes dominant.

Rare migrants on Skokholm (Numbers refer to individuals seen)

Cory's Shearwater	1	Savi's Warbler	1 (1st Welsh)
Little Egret	1	Aquatic Warbler	5 (1st Welsh)
Spoonbill	1	Great Reed Warbler	2 (1st Welsh)
Blue-winged Teal	1	Olivaceous Warbler	1 (1st British)
Red-footed Falcon	1	Icterine Warbler	11 (1st Welsh)
Spotted Crake	4	Melodius Warbler	over 20
Stone Curlew	2	Subalpine Warbler	3 (1st Welsh)
Lesser Golden Plover	1 (1st Welsh)	Sardinian Warbler	1 (1st Welsh)
			(3rd British)
Semi-palmated Sandpiper	1 (1st Welsh)	Barred Warbler	8
	(3rd British)		
Pectoral Sandpiper	10	Greenish Warbler	3
Upland Sandpiper	1 (1st Welsh)	Yellow-browed Warbler	8 (1st Welsh)
Lesser Yellowlegs	1	Radde's Warbler	1 (1st Welsh)
Sabine's Gull	6		(9th British)
Scops Owl	1	Bonelli's Warbler	1 (1st British)
Alpine Swift	1	Red-breasted Flycatcher	15
Little Swift	1 (1st Welsh)	Golden Oriole	2
	(4th British)		
Hoopoe	over 20	Woodchat Shrike	15
Wryneck	16	Red-eyed Virco	1 (1st Welsh)
			(6th British)
Short-toed Lark	(1st Welsh)	Serin	2
Richard's Pipit	over 20	Scarlet Rosefinch	5
Tawny Pipit	5 (1st Welsh)	Lapland Bunting	over 20 (1st Welsh)
Olive-backed Pipit	1 (1st British)	Ortolan Bunting	over 20
Red-throated Pipit	1 (1st Welsh)	Little Bunting	1
Bluethroat	5	Rustic Bunting	2 (1st Welsh)
Pied Wheatear	1 (1st Welsh)	Black-headed Bunting	2 (1st Welsh)
	(4th British)	Rose-breasted Grosbeak	1 (1st Welsh)
Swainson's Thrush	1 (1st British)		(4th British)
		Baltimore Oriole	1 (1st Welsh)
			(5th British)

Inland the effect of the large gull colonies is considerable, for few plants can tolerate the high nitrogen content from heavy manuring: those that can include common and sheep's sorrel, common chickweed, scarlet pimpernel and ragwort. The bracken areas are considerable, but do provide shelter for species like ground ivy, wood sage and lesser burdock and, above all others, the bluebell. Indeed, this really is the surprise of the larger islands — the vast areas of bluebells which one encounters in May on the eastern sidings of Ramsey and over much of the interior of both Skokholm and Skomer.

Grassland areas mostly comprise heavily grazed common bent and Yorkshire fog. These species have to some extent invaded the heather which is now much reduced in area, the result of salt damage following periods of drought, recovery having been impaired by rabbit and, in earlier years, grazing by goats. In all but the driest years both the North and East Ponds retain small areas of open water even during high summer, while three other areas hold water from the autumn through to late spring. The ponds, and the surrounding bogs, support an extensive flora of which silverweed, lesser spearwort, marsh pennywort, bog pimpernel and purple moor-grass are typical species. More unusual are plants such as lesser marshwort, marsh St John's-wort, shoreweed, allseed and lesser skullcap.

Skokholm, like Skomer, is renowned for its seabird colonies. The main difference between the two islands is the extremely small number of guillemots and the absence of kittiwakes from Skokholm. This is entirely due to the nature of the sandstone cliffs, the inclination of which provides few suitable sites for ledge nesting species.

Ever since the 1930s Skokholm has also been noted for the opportunities which it presents for the study of bird migration and the added excitement of occasional rare visitors from eastern Europe, Siberia and even North America. A glance at the Table will reveal the island's contribution to the British List.

From mid-March the first migrants to be seen are black redstart, wheatear, ring ouzel and chiffchaff, to be followed at the end of the month by sand martin, swallow and willow warbler. As one moves into April, the frequency of migrants increases and other species like whitethroat, sedge warbler and grasshopper warbler occur. Spring migration ceases during May, when turtle dove and swift are among the last to arrive.

Not many weeks will elapse before the spring stragglers are replaced by the first warblers on their return journey southwards to beyond the Sahara. By the end of August pied and spotted flycatchers, redstarts and whinchats, will be among those seeking a brief respite from their journey around the buildings or in the few sheltered hollows. At about the same time robins begin to arrive, and some will take up winter quarters at least until next March. During September migration continues; on occasions large passages of martins and swallows take place, most birds heading south-east. In late October and November the movement of finches, larks, pipits, thrushes and starlings can be spectacular. Most will be moving westwards and north-west, presumably heading for Ireland.

For those fortunate to live for a time on one of the islands, there is one further group of birds which arouses considerable interest —- the species which may be widespread on the mainland a few miles away but extremely rare on the island. On Skokholm these include members of the tit family, the treecreeper, magpie, rook and house sparrow. The bullfinch has only been recorded once and there are no records at all of the woodpeckers and nuthatch.

By late summer the rabbit population of Skokholm may rise to about 10,000 animals, though during the winter numbers will be halved, much more so in severe years. Rabbits were first mentioned in connection with the island as long ago as 1324, when the rabbit profits for the island of 'Skalmeye', 'Skokholm' and 'Middelholm' were £14 5s 0d. Rabbit trapping continued on the island right up to the present century, and for long periods was a major source of income to owners and tenants.

During the period of the Lockley family's occupation, efforts were made to reduce, even exterminate, the rabbits by intensive trapping and the use of cyanogas; at one time numbers were down to 400 and within sight of success when the war intervened. Several years previous, from 1936 to 1938, rabbits inoculated with the myxomatosis virus were released, but failed to infect the wild population for reasons at that time not known. Subsequently it was discovered that the rabbit flea is a vector in the transmission of the virus, and for some reason there are no rabbit fleas on Skokholm. The Skokholm rabbit population remains unaffected by myxomatosis to this day.

The only other mammal present is the house mouse, which apparently reached the island from Dale in the 1890s, in sacks carried by rabbit catchers, or so it is claimed. They quickly colonised the whole island, though the cliff slopes seem to be the main areas. Here they probably forage among the gull colonies and roosts, seeking food in the pellets of undigested food regurgitated by the gulls. By early autumn the population rises to about 5,000 and then spreads throughout the entire island, even into the buildings, despite efforts by the wardens. In the century during which they have been present, the mice have already changed in some respects from mainland animals; on average they are 14% heavier, 5% longer from head to base of tail, and their tail, relative to head and body length, is 8% longer.

'See my child how remarkable is the Island of Skomer! Come with me, I will show you every secret of its outer shore', wrote Hilaire Belloc in that classic of the sea, *The Cruise of the Nona*. Belloc, who called at **Skomer** in July 1914, was not the first yachtsman adventurer to have moored there. In 1869, E.E. Middleton in the yacht *Kate* had passed this way *en route* from Milford Haven to Ireland, and went ashore during stormy weather to be entertained by Mr Vaughan Palmer Davies. In his book *Cruise of the Kate* he briefly mentions the seabird colonies, and describes the puffin as a 'pugnacious, pompous, stupid, selfish, brutally quarrelsome dirty little bird'.

Skomer certainly is remarkable. It has a long history; the few flint flakes found suggest that early man was present soon after the last Ice Age retreated, and before a major change in sea level between 8,000 and 4,000 BC submerged the land bridge. There is numerous evidence of later arrivals in the form of field boundaries, hut sites and possibly even burial cairns, indicating that a community of up to 200 persons may have been present during the Iron Age, 600 BC to AD 100. Recent investigations by the Department of Archeology, University College, Cardiff, have revealed the possibility of residents in the Bronze Age some 1,000 years earlier.

In the latter part of the 19th century it was not only yachtsmen who were becoming aware of the splendour of Skomer. There were excursion trips from Neyland in the steamer *Malakhoff*. There were visits by naturalists who took full board at the farmhouse. Murray A. Matthews, chronicler of Pembrokeshire's bird life, records that on one visit he observed an albino puffin and regretted he had left his gun behind. An earlier albino had not been so fortunate, for that resided in a glass case in the farmhouse dining room. Robert Drane, a Cardiff chemist, of great individuality and some eccentricity, was a regular visitor, and accounts of his observations were published by the Cardiff Naturalists' Society. He was an astute observer, and the first person to recognise that the bank voles on Skomer differed in a number of respects from those on the mainland. He considered it a separate species, but later workers with much more information to hand have relegated it to an island race or sub-species of the mainland bank vole with which it will interbreed, the offspring being fully fertile. Studies in recent years have shown it to be numerous, its main habitat the dense stands of bracken.

A companion of Robert Drane on some of his visits to Pembrokeshire was a Cardiff trawler owner, J.J. Neale, who took the lease of Skomer in 1905. His concern for the wellbeing of the island and its wildlife was such that, in 1909, he had printed a statement explaining that he considered it necessary to close the island to visitors in the interests of wildlife.

Farming activities diminished as the present century unfolded, part of a trend which had begun in the mid-1880s, and which reached a peak of depression between the wars. One last gallant effort was made by Reuben Codd to farm on Skomer in the late 1940s, but this dwindled and, save for a few more years of sheep grazing, centuries of agricultural activity came to an end. Now all that remains are the crumbling walls, the empty barn where tired migrants seek shelter, the horse course, even a rusting Fordson tractor, to remind us of the once prosperous agricultural days.

The demise of agriculture has been balanced by the growth of interest of naturalists. In 1946 the West Wales Field Society, in a magnificent pioneering venture at a time of shortages and food rationing, leased Skomer for a year and established a temporary Field Centre. The story of this year and the Island's wildlife is graphically told in *Island of Skomer* by J. Buxton and R.M. Lockley. Now such centres are well established throughout the country, and thousands of visitors use those at Malham, Flatford Mill, Slapton Ley, Juniper Hall and Orielton.

In 1959 the island was purchased by the Nature Conservancy, now the Nature Conservancy Council, nearly half the cost being met as the result of an appeal by the West Wales Field Society, who subsequently leased the island from the Conservancy. These developments were unfortunately not without controversy, for a new house was built for the warden above North Haven where providentially there was an excellent spring. The old farm, badly damaged in a storm in 1954, was not to be repaired, though there are now current plans to convert part of the barn to an interpretative centre, while another outbuilding has long provided spartan accommodation for voluntary wardens and members of the Trust.

A National Nature Reserve, Skomer is renowned for its flora, its seabird colonies, its grey seals, and of course the unique Skomer vole. In addition to the large number of day visitors who make the crossing each summer from Martins Haven, smaller numbers use the simple accommodation, while each year more requests are received to make use of the island's opportunities for original research.

Skomer is the largest of the Pembrokeshire islands, being some 294 hectares in extent. East to west it measures three kilometres, north to south, two kilometres. It comprises mainly a plateau of some 60m, rising to 78m at the centre. Most striking are the bays of North and South Haven along the lines of major geological fault in the otherwise durable volcanic rock, and separated by a narrow isthmus, the centre of the largest puffin colony on the island.

The coastline is varied. At places cliffs, in others steep rock or boulder-strewn slopes. In sheltered areas, or where the slope is gentle, vegetation reaches to the splash zone and, on the northern side, there are wide terraces on the cliffs. Although not as endowed as Ramsey, there are several beaches and sea caves, most dramatic being The Lantern at the extreme east of the island. Here the cave has formed a natural arch, having three entrances, while a fourth arm extends westwards into the island, ending in a tiny sandy beach.

The interior of Skomer has a moorland quality, and at the centre one is far enough from the sea to almost lose the feeling of being on an island. There are three shallow valleys, each divided from the other areas of higher ground, dominated by rock ridges running east to west. Two major streams, a number of springs and flushes, and several man-made ponds, add further variety to the landscape. The streams and ponds dry out to varying extents during the summer months.

A glance at the table of breeding birds quickly reveals the wealth of the Skomer avifauna. Pride of place without question must go to the Manx shearwater, with about 100,000 pairs at the largest colony in the North Atlantic. Day visitors are unlikely to see them, for the birds are nocturnal, spending the day deep underground, or out at sea. Wherever there are burrows, from the cliff edge to the very centre of the island, shearwaters will be nesting, but to observe them you must spend a night or more on the island, a most exciting experience. As darkness falls, the sound from the gull colonies diminishes, though it never dies completely away. Distant lighthouses stab the sky with their beams, while on the horizon ships twinkle like slow-moving stars. Better though that such features cannot be seen, and that the night is pitch dark, the air heavy with drizzle, for then you really encounter shearwaters at their best. As the night becomes really dark it is not long before there is a swish of wings overhead, sounds of bodies moving through the vegetation, and above all, an increasing crescendo of calling, both from birds flighting in, and those on or even underground. The calls, a mixture of raucous screams, gurgles and chuckles delivered in a most exuberant manner, adds to the urgent atmosphere of the night. Walk along the main track eastwards from the old farm, and within a short distance you will see dozens of birds at close quarters in the beam of the torch, just one tiny section of the massive colony.

Most of what we know about the life history of the Manx shearwater has been gained from studies carried out first by R.M. Lockley on Skokholm and later by other investigators. Work was later to switch to Skomer under the direction of Dr C.M. Perrins of the Edward Grey Institute of the University of Oxford. Many fascinating details have been revealed, though other secrets still remain: the lengthy breeding season with the single egg being incubated for some 51 days, the ten week fedgling period for the chicks deep underground nesting in the burrow, followed by an immediate flight of 9,000 kilometres to winter quarters off the east coast of southern South America. This will be but the first of many such journeys, for shearwaters are long-lived birds, the record being of one on Skokholm which ringing showed was at least 29 years of age when last seen.

Sharing the island plateau and cliff slopes with the Manx shearwater are large colonies of three species of gulls, whose fortunes have changed dramatically over the past 50 years, as the table shows. In the early 1960s the number of great black-backed gulls on Skomer, though small, was such that their predation on Manx shearwaters and puffins was considered detrimental. The whole island was littered with corpses, often only the wings remaining, and one estimate put the annual death toll at 20,000. Control measures which commenced in 1960 slowly reduced the number of gulls though now, with the benefit of hindsight, one wonders if they would have declined without human interference. Herring gulls likewise have undergone major changes; at one time much concern was expressed regarding the effect they were having on the auk colonies. Now as elsewhere, their population has taken a dramatic downward trend.

The lesser black-backed gull on the other hand seems to suffer no inhibitions at the present. Most feed out at sea while smaller numbers frequent farmland, to search for invertebrates when fields are cleared at early potato time or at harvest. With apparently ample food, together with safe nesting areas, numbers have increased dramatically. Large sections of both Skokholm and Skomer have been taken over by these gulls, and the trend in numbers still seems upward. There is no firm evidence that the lesser black-backed gull is having an effect on other species though, if numbers continue to increase, areas used for breeding, notably by short-eared owl and curlew, could be invaded. There is also the aesthetic aspect to be considered; should the islands be allowed to develop into one huge noisy gull colony?

Skomer has the richest fauna of any of the islands, possibly of any island of a similar size around Great Britain and Ireland. The Skomer vole occupies the areas of dense bracken and, in the absence of mammalian predators, has high population densities which may reach on occasions eight times that of its mainland relative.

The open fields, even the cliff slopes in the most exposed situations, are home to the agile wood mouse, not perhaps sufficiently different to be given the prefix Skomer, but already showing divergence from the mainland form. Chief among these is its larger size, larger litters, and the extent of the yellow chest spot. The litters are especially interesting as, in the Skomer vole, litter size is marginally smaller than mainland voles and this is typical of island dwelling rodents; the wood mouse fails to follow this pattern. The other mammals are the common and pygmy shrews and of course the rabbit, the summer population of which may reach 15,000 animals. About one per cent of these are black with occasional white, or long haired, and even 'Dutch' marked animals are seen. The rabbits suffer from myxomatosis, though it was in 1965 that the last major pandemic occurred. There are no snakes on any of the Pembrokeshire islands, but on Skomer there are slow worms and lizards. Amphibians are well represented by palmate newts, frogs and enormous numbers of toads. These animals have not been looked at in detail, but no doubt exciting discoveries remain to be made. Who knows, there may even be a Skomer toad?

It was only in the 1960s that exploration of the underwater world around Skomer commenced, as equipment became readily available for the amateur diver. Soon it became apparent that there were equally important plants and animal communities beneath the waves, as existed on the cliffs and plateau above. There were pressures too, mainly the taking of large numbers of scallops, sea urchins and sea fans, together with disturbances to seabird colonies and breeding seals from divers' boats operating close beneath the cliffs. These problems brought together several organisations, both conservationalists and divers in 1970, and out of this grew a Steering Committee; in 1976 a management plan was published for a voluntary Marine Nature Reserve, one of the first in Great Britain.

The Reserve, which extends to 500m offshore, includes the whole of the perimeter of Skomer and the western end of the Marloes Peninsula. The organisations which support the Reserve are the British Sub-Aqua Club, Field Studies Council, Marloes Community Council, Milford Haven Harbour Users, Nature Conservancy Council, National Trust, Pembrokeshire Coast National Park Authority, Preseli District Council, South Wales Sea Fisheries Committee, Welsh Association of Sub-Aqua Clubs, Welsh Anglers Council, Welsh Water Authority and West Wales Trust for Nature Conservation.

The topography of Skomer, with the sounds separating it from the mainland, has resulted in a superb range of underwater habitats, from soft sands and muds on the sheltered north side, to areas of extreme exposure to wave actions and fierce tidal streams. This variety of habitats, and the communities and species which they support, make the marine reserve area of such high conservation and scientific interest. A number of the plants and animals are of international, national and regional importance, including several species of seaweed, sea fingers, sea fan and corals.

In 1981 the opportunity presented by the Wildlife and Countryside Act for the establishment of Marine Nature Reserves was welcomed. One major result would be the provision of Government funding, where previously there had been but a shoe string budget. Unfortunately progress over Skomer has been slow, and at the time of writing in the spring of 1986, consultations were still in progress. It is hoped that by the end of the year or early in 1987 these will be completed, and Skomer will take its rightful place as one of the first statutory Marine Nature Reserves.

Skomer Vole

ABOVE: Iron Age storage hut site in North Stream valley, Skomer. (DRS)
BELOW: The ruined farm and outbuildings on Skomer. (NCC)

ABOVE: Sheep dipping in the North Stream valley on Skomer in the 1880s. (HB) BELOW: Reuben Codd, the last man to farm Skomer, with Prince the horse and visitors' luggage. (RH)

ABOVE: The Skomer boat service in 1946. (WWTNC) BELOW: Pigstone Bay, Skomer, the most westerly point on the island, and a regular place to observe seals. (NCC)

ABOVE: The gannet colony on Grassholm. (BT) BELOW LEFT: Adult gannets. (MA) RIGHT: Juvenile gannet at the time of departure from the colony. (JP)

ABOVE LEFT: Adult shag with just a trace of the breeding season crest remaining. (FP) RIGHT: Kittiwake colony; the bulky nests may clearly be seen. (JP) BELOW: Manx shearwater, most numerous of all the seabirds breeding on the Pembrokeshire islands. (JT)

Natural History Highlights

Polecat

Flora: Carmarthenshire is quite a rich county botanically, containing a diversity of habitats, but has a reputation, chiefly fostered by past botanists, of being rather dull, a reputation that has been difficult to shake off. The county can probably claim to be the stronghold, in a global sense, of whorled caraway. This species is frequent in damp, usually acidic meadows, and is particularly abundant on the coalfield, where the continuance of traditional farming practices has favoured its survival. Associated species often include meadow thistle, bogbean, saw-wort and marsh orchids. Several species of sedges and rushes are also often present and, in the richest sites, petty whin, dyer's green-weed and marsh valerian occur.

The upland, northern half of the county still supports relatively large areas of hanging sessile oak woodland, a habitat principally confined to western Britain. Several montane ferns are occasionally found on the rocky crags in this district, and mountain pansy reaches its south-western limit in Britain hereabouts.

The Old Red Sandstone outcrop is mostly intensively farmed, but in the east, where it forms the elevated, base-rich cliffs around Llyn-y-Fan-Fach, several arctic-alpine species reach their southern limits. These include dwarf willow and northern bedstraw. Other plants which reach their southern limits in the county are water sedge at Talley, and bog-rosemary at Beacon and Llanllwch Bogs, although there are two old records for the latter from Somerset.

The Carboniferous Limestone outcrop to the east of Llandybie forms part of Mynydd Du and is heavily sheep grazed. Much of the faulted outlier of Carreg Cennen, however, is so precipitous as to be inaccessible to grazing animals, and chives and western spiked speedwell are to be found on the rock. Further west, native woodland surmounts the limestone outcrop, but many active quarries are rapidly destroying the few areas remaining. Herb paris and toothwort, both at or near their western limits in England and Wales are, however, occasionally encountered.

The calcareous coastal dune systems support a rich flora, including specialities such as fen orchid and dune gentian. There are extensive salt marshes, some of the best in western Britain, which include sites for bulbous foxtail. Coastal fens and freshwater habitats support species such as marsh pea and frogbit.

The doubtfully native Tenby daffodil is frequent in the western part of the county, particularly on roadsides, although also in a few riverside woods. Another doubtful native in the county, prostrate toadflax, is found at several old railway sites in the Pembrey area.

With the initiation of intensive recording for a forthcoming Flora of Carmarthenshire, several species thought to be extinct in the county by R.F. May in his list published in 1967, have been refound. Notable among these are hay-scented buckler fern, dwarf chickweed, roseroot, great broomrape, meadow clary, and ivy-leaved duckweed. Several new discoveries have been made, including bog orchid and greater spearwort, but there appears still to be much scope for further research and no doubt much excitement for botanists for many years to come.

Ceredigion is a predominantly acidic district with no high mountains, and its flora is not outstanding for rarities, but there are many compensations for the botanical visitor. Several highly attractive species with a western Atlantic distribution, uncommon in Britain as a whole, can be seen here in profusion, notably spring squill on the clifftops around Gwbert and Mwnt, wavy St John's-wort in some of the valley mires such as the Rhos-y-fforest nature reserve, whorled caraway on some of the rhosydd such as the Rhos Glyn-yr-helyg nature reserve, ivy-leaved bellflower on the slopes of the Llyfnant and Cwm Ystwyth and other valleys, and wood bitter-vetch on the margins of some of the few remaining upland hay meadows. This last species has declined drastically in recent decades as land use has changed, and other species have been lost or nearly lost for the same reason. Destruction of heathlands and improvement of most of the upland pastures have meant that mountain everlasting, fairly frequent fifty years ago, has not been seen since 1960, and field gentian has not been reliably reported for even longer. Bastard balm, known from what is now the Coed Cwm-du nature reserve in the 1940s and from a wood near Brynhoffnant in the 1950s, has been lost probably because of the decline of woodland management. A few arable weeds have also become extinct, but a surprising number of rare weeds such as annual knawel, field madder and dwarf spurge persist in the arable fields between Aberporth and Gwbert.

Rare Pembrokeshire Plants - Numbers within Broad Habitats

	Red Data Book Species	Other Nationally Scarce Species	Local Rare Species	Totals
Coast (Cliff species in brackets)	9 (7)	41 (22)	43 (21)	93 (50)
Wetland	0	4	29	33
Lowland Heath	1	9	9	19
Woodland, Scrub, Hedgebank, Verge	0	2	15	17
Disturbed Ground, Quarries, Walls, Railways	0	0	17	17
Grassland	1	0	6	7
Totals	11	56	119	186

Definition

1. Red Data Book species:- less than sixteen 10 kilometre squares in Britain. 2. Other nationally scarce species:- between 16 - 100 ten kilometre squares in Britain 3. Local Rare Species:- 3 or less 10 kilometre squares in Pembrokeshire.

If the list of extinctions is small compared with most of Britain and even with much of Wales, the opportunity for making new discoveries still exists, as many of the more characteristic habitats of the district still survive, even if they are reduced in size. J.H. Salter, the author of *The Flowering Plants and Ferns of Cardiganshire*, said that it would be a happy day for him if he could meet with any one of three particular species that had eluded discovery in the county: bog orchid, yellow bartsia, and pale butterwort. Since Salter's day, fortunate botanists have discovered the first two, and pale butterwort may well occur on one of the remaining rhosydd or boggy heaths in the south-west of the county, not far from its Welsh stronghold across the Teifi on Mynydd Preseli. Lanceolate spleenwort, which occurs in a number of localities in all the other old counties around Cardigan Bay, is another species surprisingly still unrecorded from Ceredigion.

The absence of strongly calcareous rocks in the uplands explains the absence of several plants familiar elsewhere in Wales, such as purple saxifrage and roseroot; although one plant of the lime-loving mossy saxifrage has been known on Plynlimon for over 80 years, a second patch of it has vanished in recent years. Conversely, parsley fern, a plant of acid unstable screes that might be expected to be abundant in Ceredigion, is represented by only four tufts; it is probably discouraged by the enriching effects of the salt-laden south-west winds.

Four species, forked spleenwort, lesser clubmoss in the Tywi valley and in Cwm Ystwyth starry saxifrage abundant in places on Plynlimon and in the Rheidol gorge, and awlwort in the Llyn Fanod nature reserve, reach their southernmost limit in Britain in Ceredigion. Since the disappearance of bastard balm, just one species, Cornish moneywort, which has been recorded in several places near Llandysul and still occurs in one wet wood, reaches its northern limit in Britain here.

The old county of Pembrokeshire, with its pronounced marine plateaux, is the most westerly extremity of Wales. No part is far from the sea, and nowhere escapes the impact of stormy sea winds. Along the predominantly cliff coast, headlands vary from Carboniferous Limestone to igneous rocks of precambrian age. Inland, apart from a few tors near the summits of gently undulating hills, the rocks lie buried under mantles of clay and sometimes under fluvio-glacial sands. The flora of some 830 native species has many affinities with Cornwall — another truly maritime province.

Montane plants are virtually absent. Mynydd Preseli reaches only 541m — not high enough to have escaped tree cover when the climate was warmer some 6,000 years ago. The coastal fringes, whether salt marsh, sand dunes or the major cliffs, probably remained open throughout the period of maximum tree cover. The principal headlands with adjacent deep water provided the greatest continuity of open conditions, and it is among these natural rock gardens that the botanist will be amply rewarded. Not only are there abundant sheets of thrift and sea campion in quantities never encountered in the mountains of Wales, but there are numerous frost-sensitive species seldom seen far inland. Of the 11 British Red Data Book plants found in Pembrokeshire, nine are coastal, seven being confined to sea-cliffs. The hairy greenweed has about half of its British localities on the maritime heaths of the cliffs between Newgale and Fishguard. Wild asparagus persists, but only just, on the limestone cliffs near Tenby, where it was described as plentiful in 1848 in Dr R.W. Falconer's plant list for Tenby. A recent find is goldilocks aster from the maritime grassland and heath on the limestone cliffs of Castlemartin, where more populations of this late summer flower undoubtedly await discovery. These three species are all creeping perennials, able to survive the frequent, salt-laden gales.

Even more salt tolerant are the rock sea-lavenders for which Pembrokeshire is notorious — two endemic British species have their strongholds on the cliffs of Pembrokeshire: *Limonium paradoxum* and *L. transwallianum*. The small restharrow is the only annual of the Pembrokeshire Red Data Book species and its irregular flowering and open, south-facing limestone haunts seem appropriate for a species of the Mediterranean and Middle East. The perennial centaury flourishes in maritime heath and grassland near Newport. This — its most northerly station in Europe — was where it was first discovered in Britain, by T.B. Rhys, a local headmaster, on 16 August 1918. Another cliff plant at its most northerly location is hairy bird's-foot-trefoil which occurs near Dale and St Davids. In complete contrast is the arctic-alpine roseroot which hugs a steep north facing crag on St Davids Head; its most southerly coastal site in Britain.

North Pembrokeshire has the finest lowland wet heaths in Wales. Unlike their maritime equivalents, these are products of the burning and grazing activities of early agriculturalists whose monuments litter the landscape. These lowland heaths in the St Davids area share plants with the nearby cliff heaths; yellow centaury, three-lobed water-crowfoot and pale heath violet. Interestingly, sea plantain is frequently mixed in with these notable plants, even on the inland heaths. Further away from the coast around the lower slopes of Mynydd Preseli, extensive sheep-grazed heathland harbours the largest Welsh population of marsh clubmoss. Other plants of the peaty open areas include four insectivorous species: pale butterwort, common butterwort, long-leaved sundew and round-leaved sundew. The widespread flushes and small basin bogs have populations of dioecious sedge and the diminutive bog orchid.

The inland heaths of St Davids grade into sedge fens and swamp of bogbean towards their wetter centres. The presence of the rare fibrous tussock sedge at the margins of these swamps is indicative of the Irish Connection in the Pembrokeshire flora. This species' nearest population lies across the Irish Sea in the fens of Eire. Similar sedge fens are still frequent at the coastal ends of the many valleys that were over-deepened by glacial meltwater during the Ice Ages. Greater tussock sedge, often forming stumps over six feet high, is a feature of the Pembrokeshire fens, where it is usually mixed with greater and lesser pond sedges. Botanists encountering stands of the tussock sedge acquire the knack of leaping from tussock to tussock, so enchantingly described by Philip Gosse as early as 1856 during an outing to Penally marsh, Tenby. Calcareous fens like that at Penally and another at Castlemartin Corse have much in common with those in Anglesey, with species like marsh fern, fen pondweed, greater spearwort, and the recently discovered common meadow rue. Other notables include tufted and slender sedge, while the graceful galingale still survives in one large clump at Penally and great fen-sedge persists in the inland Llangloffan fen.

Although Pembrokeshire carries little semi-natural woodland, its hedgebanks and verges are exceptional for their show of early flowers. Many of these are woodland species like the bluebell and red campion. Scarce plants of woodland origin are Soloman's-seal, frequent on road verges near the Carmarthen border, and much more rarely, the equally attractive, bastard balm. Coastal plants like common scurvygrass have recently spread inland along the highways. Its scented floral outburst in late April, along with the overpowering heavy perfume of gorse, is the harbinger of spring. The scurvygrass is followed by primroses of both pink and yellow colours and then by cowslips. Once abundant in the fields, they are now largely confined to roadsides and coastal grasslands, the verges of the A40 road outstanding for their banks of cowslips in 1985. Along with verges, churchyards are the last significant inland refuges for the many colourful plants of dry, unimproved meadows. Slebech and Nevern churchyards are especially fine examples. The Tenby daffodil, a Red Data Book plant of obscure origin, can still be seen on verges as well as in cottage gardens. Plants of dry disturbed ground such as corn parsley and pale flax can also be found on roadsides, while the walls and surrounds of the many historical monuments and old buildings have additional species like spotted medick, fiddle dock and the handsome compact brome grass.

No summary of the Pembrokeshire flora would be complete without mention of its lichens. The clean air and oceanic climate both favour lichens and the woods and cliffs are renowned for their lichen communities. Over 300 different species are found in one damp rocky oakwood at Tycanol in the north, Stackpole in the south boasts of ground lichens of international quality, and the islands of Skomer and Skokholm are also of similar value for their species of cliff and rock outcrops. Much further recording of these lichens and also of mosses is needed. Even with the flowering plants, native species can still be discovered — over ten new ones have been identified since 1980! The importance of careful recording cannot be overemphasised. Local botanists are especially indebted to the pioneering work of the late Tommie Warren Davis in this respect. Both he and Bertram Lloyd, who visited Pembrokeshire regularly in the 1920s and 1930s, maintained detailed diaries of all their excursions. Their diary entries delight the natural history reader by taking him on plant hunting journeys through the unrivalled variety of West Wales.

Birds: At present, the Carmarthenshire bird list total numbers 277, made up of 83 resident breeders, 27 summer visitors who breed, and an additional 15 other species who have bred during this century. Over the last thirty years gains among breeding birds have been great crested grebe, fulmar, goosander, feral greylag and Canada geese, goshawk, collared dove, reed warbler, siskin and crossbill. Those lost have been shoveler, red-leg and grey partridge (apart from isolated releases), black-headed gull, woodlark, nightingale, corn bunting and oyster-catcher, while one or two pairs of teal, golden plover, redshank, lesser black-backed gull, long-eared owl, short-eared owl have been known to breed irregularly. Merlin, red grouse, black grouse and woodcock have become even more restricted than formerly in the northern uplands, while corncrake, barn owl, turtle dove, nightjar and yellow wagtail have declined in the lowlands, and herring gull on the coast. A pair of little ringed plover appeared to set up territory on a shingle bank on the Tywi, near Llandeilo in June 1984, but were presumed to have been prevented from nesting due to disturbance by cattle. Other possible breeders in the near future may be red-breasted merganser, ruddy duck and marsh harrier.

The areas of undulating uplands and moorlands with their narrow, steep-sided, wooded valleys, hold a good cross-section of the birds which favour such habitats, including pied flycatchers, redstarts, wood warblers and tree pipits in summer, with resident ravens, buzzards and red kites. The extensive plantations of non-native conifers are not as interesting as the smaller broad-leaved woodlands, but some have attracted new breeding residents in redpoll, siskin, crossbill and one or two pairs of goshawks. Winter wildfowl, including mallard, teal, widgeon, lapwing, golden plover and curlew, together with a flock of Siberian white-fronted geese, occur in good numbers along the Tywi water meadows in the vicinity of Dryslwyn, the most important inland wintering ground in Dyfed.

The three estuaries and northern bank of the Burry Inlet attract a variety of waders, wildfowl and other species, in season. The sea marshes and sandy burrows between Cydweli and Pwll have produced in recent years exciting rare vagrants, including cream-coloured courser, white-rumped, buff-breasted, Baird's and pectoral sandpiper and killdeer plover. The Witchett Pool, sited amongst the sand dunes of Laugharne Burrows, as well as attracting a large variety of species, including shoveler, gadwall, scaup and bittern, has breeding grebes, pochard, tufted duck and greylag geese, with ruddy duck and marsh harrier as possible future colonisers. Large flocks of common scoter occur offshore on Carmarthen Bay, while a small number of herring gull and fulmar breed on the cliffs, west of Pendine. Compared to the south-east and Tywi valley, the north-western sector and Teifi valley has been rather neglected by ornithologists, despite the existence of many interesting habitats worthy of our attention.

Ceredigion has a full range of bird habitats, from montane heaths and crags through wet moorland, commercial forest and native woodland, marginal and productive farmland, lakes and rivers, valley bogs, coastal marshes, sand-dunes, estuaries and seacliffs, to the open sea. Human settlements are small and scattered. The countryside is changing, but less drastically than elsewhere. In this landscape 114 bird species have nested frequently in the past decade, and a further 25 bred formerly or occasionally in this century. The total county list stands at 267 wild species.

Notable features of the avifauna include especially good representation of the typical birds of open uplands and upland native woodland, with red and black grouse, merlin, peregrine, golden plover, dunlin, raven, buzzard, red kite, tree pipit, ring ouzel, redstart, wood warbler, pied flycatcher, and many others. Wetland birds abound: Ceredigion probably has more breeding teal than any other Welsh county, and the heron population is notably high. There are varied seabird communities, including two of the largest cormorant colonies in Wales, and four-figure colonies of auks and gulls. Choughs and stonechats are found along the coast, and locally inland. Winter wildfowl occur in good numbers, and include a small flock of Greenland whitefronted geese and the most southerly regular group of whooper swans. Passage waders are an important feature, especially at the Dyfi, though restricted recently by the spread of cord-grass over the mudflats. Productive waters off the Dyfi mouth hold in season some of the largest gatherings of red-throated divers, great crested grebes, and sandwich terns recorded in Wales, and attract a wide variety of other seabirds.

The status of some species has radically changed this century. The disappearance of once regular breeders such as turtle dove, wood-lark, yellow wagtail, red-backed shrike, cirl and corn buntings is presumably due to climatic change, while corncrake, little tern, and puffin perhaps departed mainly by human agency. The grey partridge population depends on released birds, though it formerly flourished in thousands. The nightjar seems to be on the edge of extinction. Other birds have declined, perhaps mainly through loss of their specialised habitat, among them merlin, red grouse, golden plover, lapwing, ringed plover, snipe, barn owl, song thrush and yellowhammer. Others are probably only temporarily depressed: victims of the Sahel drought like sand martin, sedge warbler, and whitethroat.

The gains include little owl, fulmar and, more recently, great crested grebe, red-breasted merganser, goosander, kittiwake, collared dove, reed warbler, and siskin; most seem well established. Manx shearwater, Canada goose, tree sparrow, and crossbill have all nested for the first time, and could persist. Increasing species included nuthatch and starling earlier this century, while black grouse, lesser black-backed gull, guillemot, razorbill, pied flycatcher and probably lesser whitethroat have been on an upward trend lately. Red kite and peregrine staged a splendid come-back after near-extinction. Herring gulls multiplied but now seem to be in decline, probably due to outbreaks of botulism. A few other species appear more often lately and may eventually breed; tufted duck, pochard, goshawk, hobby, and black guillemot are among the candidates.

North Ceredigion has always been better watched than the south, due mainly to successive generations of bird watchers among the students and professional people in Aberystwyth, and more recently the employees of conservation bodies based in the north. There is still much to be learned about the distribution and status of even common birds in the south, especially inland.

Such a variety of habitat occurs within Pembrokeshire that it can be no surprise that the avifaunal list with 318 species,with a few still under consideration, is one of the richest in Wales.Of this total about a third breed or have bred. There have, however, been losses; for instance, cirl and corn buntings have ceased to breed since the late 1940s; woodlarks disappeared as a result of the hard winter of 1962-63 and have not yet managed to return. On the other hand gains have been registered – collared dove, for instance, in 1962; reed warbler in 1974; Cetti's warbler probably breeds, and pied flycatcher has done so since 1978.

The county is well placed geographically; this south-westerly extension of Wales, with its offshore islands and diverse coast is a first landfall for many lost migrants, both Old and New World species. Ronald M. Lockley set up Britain's first bird observatory on Skokholm in 1933 when he built a Heligoland bird trap there. The British Trust for Ornithology officially recognised this Pembrokeshire initiative which stimulated a nationwide interest in migration studies. Between 1928 and 1976 almost 170,000 Manx shearwaters were ringed on Skokholm – breeding data, movement and lifespan facts were the reward. The Observatory work is well illustrated by the following extract from the Skokholm Bird Observatory Report for 1967: 'The year proved to be a record one in almost every aspect. Skokholm once more broke its own record for numbers of birds ringed in a year, 16,616 birds of 91 species were ringed. Of this total just over 10,500 were Manx shearwaters whilst a further 2,500 were storm petrels. The year provided eight new species for the island ringing list: great reed warbler, serin, stock dove, snow bunting and four American passerines, rose-breasted grosbeak, Baltimore oriole, red-eyed vireo and olive-backed thrush.

'One hundred and fifty-one species on the island was a record for any year. Of these, great reed warbler, Richard's pipit, serin, and the four American passerines were new additions to the island list. Rare species seen included pintail, white-fronted goose, Kentish plover, pectoral sandpiper, sanderling, great skua, little gull, roseate tern, nightjar, short-toed lark, bluethroat, reed warbler, barred warbler, firecrest, red-breasted flycatcher, tawny pipit, woodchat shrike, twite, ortolan bunting, Lapland and snow buntings.'

Skokholm is no longer an observatory in the bird-ringing sense, but observation studies continue. The observatory workers, the local resident ornithologists, numerous interested holiday visitors, over the years, have added substantially to Pembrokeshire birds for, of the county total, many species are included which have occurred only once or but a few times. Vagrants like the little egret, passage visitors like the garganey which can usually be seen in spring on the St David's Peninsula, and winter visitors like the fieldfare, make up the rest of the total.

The tidal reaches and creeks of Milford Haven prove of great value for wintering waders and duck, and co-ordinated counts have revealed nationally important numbers of such species as shelduck, teal and redshank. Our major heronries are related to the Cleddau system and fluctuate according to the severity of winter periods. Llys-y-fran reservoir, some 74 hectares of water since filling in 1972, sustains large numbers of winter duck, particularly pochard and tufted, with goldeneye, mallard and teal at times. A feature of Llys-y-fran in recent years has been the flock of wintering lesser black-backed gulls roosting there, possibly a reflection of island breeding success and changing habits. Apart from Bosherston Pools, large open freshwater bodies are few, so such reservoirs are well monitored.

Wetland species like snipe and curlew have decreased, perhaps because of habitat loss and deterioration; the wet commons, for instance, are not as well grazed as in the past, and tall growth climaxing in willow scrub has doubtless rendered the sites less suitable for breeding. On farmland, intensive grazing regimes and forage cropping, silage cutting for instance, have made the habitat less safe and productive for corncrake and grey partridge, although other factors such as migrational hazards and climatic changes have taken toll.

Cuckoos and meadow pipits still duo in the upland areas, with curlew and raven ever present; wheatears breed but seemingly in lesser numbers. In winter snow buntings spend time on Preseli and adjacent summits but more often on the coast. Afforestation is not extensive on our uplands but plantations of Sitka spruce with lodge pole pine and some larch do exist and, while such areas have ousted species of the open hill, short-eared owls find the less successful plantings, the young plantations and their edges, suitable to their use; hen harriers benefit similarly, and modest roosts of this wintering species occur at plantation fringes here and there. Large winter roosts of starlings and fieldfares, with redwings, occur in these plantations, while the more mature areas are attracting siskins and redpolls, and even redstarts can be located on Mynydd Preseli these days. Whinchats and stonechats occur, the latter particularly in coastal gorse-grown habitats. A few dippers and grey wagtails frequent upland streams, and occasionally ring ouzel breeds in an upland quarry.

Our broadleaved woodlands are limited in number but redstarts, wood warblers and now even pied flycatchers can be seen in the sessile oakwoods. Tits, nuthatches, treecreepers, green and great spotted woodpeckers occur, with lesser spotted woodpecker less frequently found. Nightjars are now infrequent and one must search hard in the younger plantations to see them. Buzzard and kestrel are frequent, and red kite visit the country more than previously, no doubt reflecting the slow but definite success in neighbouring Carmarthen and Ceredigion. Merlins seem now to be only wintering birds but will the hobby come to breed with us? Records are increasing. Peregrines have not yet recovered to their pre-war status of 35 pairs but, from one or two pairs in the 1960s, some 20 or so sites are now occupied. Goshawks are often seen, and displaying birds have been noted in recent years; do they in fact breed?

If one is interested in seabird passage, visit Strumble or St David's Heads from late July to late October, early morning for preference and, with onshore winds following south-westerly systems, an astonishing species tally can be made. Sooty, great and even Cory's shearwaters are possibles, while the little shearwater has also been seen: great, Arctic, pomarine and longtailed skuas, Mediterranean and Sabine's gulls and, according to season, glaucous and Iceland gulls. Common, Arctic, Sandwich, little and even black tern pass the headlands, but alas, none breeds in Pembrokeshire. Larks, pipits, finches, starlings and thrushes occur at these headlands on autumn migration and spectacular movements may be observed.

Pembrokeshire's Organising Committee for Ornithological Research (POCOR) has initiated a five year breeding bird survey on a tetrad basis with a view to a new county avifauna, other species surveys are ongoing, but truly, the more one learns the more there seems to be learnt.

Mammals: The varied landscape of Carmarthenshire, with its wide selection of surviving semi-natural habitat, is the home of a reasonably diverse mammalian fauna and, although there have been obvious losses since early historical times, the species that are left still give enjoyment and opportunity for study to the naturalist.

The earliest records of mammals in the county could be said to be those remains found in the now destroyed cave at Coegan on the Carboniferous Limestone near Laugharne, where archaeologists at the turn of the century found woolly rhino, mammoth, reindeer, cave hyaena, cave bear, cave lion, giant ox and a primitive horse species, later mixing of the cave deposits resulting in this perhaps strange juxtaposition of cold and warm climatic indicator species.

Red deer antlers have been found in peat deposits on more than one occasion, while the roe deer was probably quite widespread and no doubt widely hunted until at least medieval times; this forest-loving deer is now extinct in the county, but the fallow deer in the Tywi Valley centred on Dinefwr are a later introduction. The brown bear, wolf, wild boar, pine marten and wild cat are long extinct, but it is interesting to note the record of Geraldus Cambrensis of beavers on the Teifi sometime prior to c1200.

In spite of the above catalogue of extinctions, many species still thrive in the county. The polecat has spread in recent years from its traditional strongholds in the north, and it now occurs even on the margins of towns, and sometimes enters the suburbs – as roadside victims testify. Otters have, as in many areas of Britain, declined, though they are still present with varying degrees of regularity and in low numbers on many of the county's unpolluted rivers. Small, black 'otters' that are sometimes reported, refer to the now common introduced mink which, in spite of trapping and general persecution, is increasing.

The fox is widespread, even in suburban areas and parks, while the badger or *mochyn daear* 'earth pig' is similarly common, though regrettably still subject to illegal persecution. As well as in the more typical woodland habitat, badger setts have been noted high on treeless hillsides overlooking the wooded valleys, and once a badger nonchalantly ambled across a dune-slack at Tywyn Burrows in full daylight!

The red squirrel is now rare, but it still survives around the large coniferous forests of N and NE Carmarthenshire; there have also been recent records from the Upper Tywi Valley around Rhandirmwyn. Its introduced relative, the grey squirrel, is common and widespread and, in times of high population density and during autumn dispersal, it is seen in a great variety of habitats, even open moorland and coastal dunes.

A list of fairly widespread mammals would include hedgehog, stoat, weasel, wood mouse, house mouse, brown rat, mole, short-tailed vole, common and pygmy shrew. There are only a few records of the water shrew (as cat victims), two records of the dormouse (Cwm Lliedi, Llanelli, and near Ferryside), a couple of records for the yellow-necked mouse, which include several trapped raiding an apple store at Rhandirmwyn! As yet there are no absolutely certain records for the diminutive harvest mouse. The hare is widespread in the county in open habitat, while the water vole is fairly frequent beside still or slow water, at least in the coastal areas.

The only marine mammal to be regularly seen off Carmarthenshire's coast is the grey seal, sometimes swimming up-estuary as far as the docks at Llanelli, or even many kilometres inland up the Tywi river, and once as far as Dryslwyn to the amazement of bird watchers gathered to count winter wildfowl!

Lastly, turning to reptiles and amphibians, the common lizard, frog, toad, slow worm, grass snake and adder are reasonably widespread. The common newt is mostly but not exclusively coastal; the palmate newt on the other hand is able to survive in the more acidic upland areas.

LEFT: Killdeer plover, a rare vagrant to Carmarthenshire. (DF) BELOW: Polecat, now widely distributed. (DF) RIGHT: Harvest mouse, probably more widespread than the records suggest. (DF)

Much remains to be discovered about the distribution of mammals in Ceredigion, where there are only scattered records even for widespread species. The fortunes of three carnivores are of special interest in the county and well worth re-telling.

There is much confusion concerning the authenticity of past records and references to the wild cat in Wales. Many of the records in parish accounts and other documents undoubtedly refer to feral domestic cats. Nevertheless it seems likely that the wild cat lingered on in mid-Wales until the 19th century, reduced in numbers and eventually to extinction, most probably due to trapping and hunting. Ceredigion has the doubtful distinction of being the county with the last apparently genuine record of this fine animal in Wales. In 1895 a gamekeeper shot a cat on an estate near Talybont which, on careful examination, was found to have all the characteristics of the wild cat rather than the feral variety.

The pine marten has just avoided the same fate in Wales, where there are from time to time still reports of its occurrence. In the Middle Ages and beyond its fur was highly prized, being second only to that of the beaver in value. The decline of the pine marten in Wales, like that of the wild cat, is largely due to trapping and hunting, with additionally loss of habitat. Large woodland areas next to hill land with deep heather, scree and rocky crags are probably its main requirements. Maybe the population is now so reduced and fragmented that recovery is likely to be slow if achieved at all.

At one time the polecat was found throughout Great Britain; indeed the first specimen exhibited at the London Zoo was trapped in Regents Park in 1828. Not many years were to elapse before the polecat was suffering the same fate as the wild cat and pine marten, with its numbers drastically reduced by the late 19th century. Eighty years ago it had become extinct in Scotland, most of England and large parts of Wales, where its last stronghold was the centre of the Principality, and in particular Ceredigion. Here it was to remain reasonably common; for instance, in the Clarach area no less than 66 were killed during the second half of 1930.

Trapping was the main hazard faced by polecats, for they are easily caught. Many were deliberately trapped; others were accidentally taken in gin-traps. The arrival of myxomatosis in the mid-1950s and the banning of gin-traps in 1958 greatly reduced the pressure on polecats. From then on their numbers increased with, at the same time, a rapid expansion of range, so that the polecat is now found commonly throughout Wales and into the border counties.

Of particular interest in north Ceredigion is the incidence of the so called 'red' or erythristic form of the polecat. In this the brown or black colour in the normal coat is replaced by a red, or reddish-brown tint. The first specimens seem to have been two shot on Cors Caron in 1903 and 1904 and sent to Hutchings, the Aberystwyth taxidermists. There have been a scattering of records since, together with a few from neighbouring parts of Powys and Gwynedd.

Some 33 species of mammal, excluding cetaceans, have been recorded from the vice-county of Pembrokeshire which, for three 'aquatic' species, the otter, American mink and grey seal, is their stronghold in Dyfed.

Survey work, particularly since the late 1970s, has shown that the otter is widespread in small numbers in Pembrokeshire which, together with adjoining parts of Carmarthen and Ceredigion, is its headquarters in England and Wales. The otters' secret and largely nocturnal habits mean that they are rarely seen, but their presence in an area can be detected by searching the waterways and wetlands for their footprints, or spraints or droppings. Spraints are usually black and full of tiny fish bones, and when fresh look oily and have a characteristic, and pleasant, sweet musky smell.

The amount of waterway that each otter inhabits – its home range – varies according to the animal's age and sex, to the size of the local otter population, and to a number of environmental factors, such as the amount of food and the quality of habitat available. The home range of a female otter is in the region of 16 kilometres, whereas for a male it can be as large as 40 kilometres, which is the length of the Teifi between Cardigan and Llandysul! As home ranges often overlap, a dominant male might share parts of his patch with several females and other males.

As well as a good food supply, otters also need secure places in which to rest during the day. A variety of sites are used, but the most common are in dense bankside vegetation such as bramble and blackthorn thickets, and in the root systems of mature bankside trees, especially oak, ash and sycamore. The most undisturbed and secure of these sites is chosen by the female otter to raise her litter from two to four cubs.

The highest percentage of signs of otter actually recorded in the 1977–78 survey occurred on the Gwaun, Western Cleddau, Brandy Brook and Solva catchments. Subsequently more detailed work revealed that the upper reaches of the Western Cleddau were nationally important, especially when woodland, scrub and wetland occur alongside or close to the river. Tributaries and sections of main river subjected to management, especially with follow-up improvements to adjoining land, have a significantly lower level of otter activity. Only four potential holts were discovered, and it is thought likely that, in the absence of suitable tree root sites, otters are breeding in some of the undisturbed wetland and scrub areas on the river banks.

Although there are occasional records from the shores of Milford Haven, and in the past even of animals drowned in lobster pots, surprisingly little is known as to whether otters make much use of estuaries and coastal habitats in the county.

The Vincent Wildlife Trust has provided a great deal of advice for landowners and the Welsh Water Authority regarding management schemes, which take account of the conservation requirements of otters. Otter havens have been established at several key spots; these provide quiet undisturbed habitats for this shy mammal. Log pile holts have been constructed where natural holts have been destroyed or are scarce. A number of tree planting and other schemes involving the management of bankside vegetation have been embarked upon with the co-operation of landowners. As long as clean water, plenty of food and suitable cover are available, West Wales will remain one of the strongholds of this beautiful but elusive animal.

Not of such absorbing interest as the otter, but nevertheless an animal now firmly established, is the American mink. Fur farms commenced in Great Britain in 1929 and escaped animals were recorded shortly afterwards, but it was not until 1956 that breeding was first confirmed in the wild, on the upper reaches of the River Teign in Devon. In Pembrokeshire there were a number of farms, the last one at West Hook on the Western Cleddau closing in the early 1970s. Most records date from then, and now American mink are common on all water courses in Pembrokeshire, save possibly those in the extreme south-west.

Although the Pembrokeshire islands of Ramsey and Skomer are the headquarters of the grey seal in Wales they are nevertheless a frequent sight elsewhere along the coast. Some wander regularly into estuarine waters, and occasionally follow the Milford Haven waterway right to its tidal limit at Haverfordwest weir.

Studies on the grey seal population were originally confined to the islands and a few mainland sites, but during the mid-1970s several surveys and assessments of the whole population were made. These clearly showed that the vast majority of grey seals in Wales are in Pembrokeshire, where about 660 pups are born each autumn, representing a total population of adults and young of up to 3,000 animals. Numbers are still slowly increasing, though this means that less suitable breeding beaches where pup mortality is high have to be used.

At least 11 species of cetacean have been recorded off the coast of Pembrokeshire. In 1913 the systematic recording of cetaceans stranded on the British coast was commenced by the Keeper of Zoology at the British Museum (Natural History), Sir Sidney F. Harmer. Whales, dolphins and the porpoise belong to the Crown or to certain Lords of the Manor, and all those stranded have to be reported to the Receivers of Wreck or the Coastguard. This has resulted in a good deal of information being acquired as to the occurrence of these animals in British waters. One of the most noteworthy strandings was the 22 foot bottle-nosed whale at Fishguard in 1936. This was sent by road to the Natural History Museum and aroused much interest at the time.

In March 1928 a vast school of many hundreds, if not a thousand, porpoises entered Dale roads, an observer noting that 'the sea from close inshore was boiling with surfacing and leaping porpoises, an astounding sight'. Although porpoises are the most regularly seen of all cetaceans off the Pembrokeshire coast, the numbers are now usually extremely small.

There have been two records of bottle-nosed dolphins which have created a great deal of interest. In April 1975 one 'adopted' the Skomer boat as it made its way around St Ann's Head *en route* to Martinshaven for the first Skomer visitors. The dolphin took up station at Martinshaven throughout the season, occasionally following the boat to Skomer, Skokholm and even back to Dale. It vanished as soon as the boat ceased the island service during mid-September. During 1984 and up to the early autumn of 1985 another bottle-nosed dolphin adopted Solva and the many boats there, giving much delight to hundreds of visitors, a few of which even swam with it in the harbour. Alas, it too disappeared as mysteriously as it had arrived.

Sightings of cetaceans offshore are not infrequent, especially on headlands like Strumble, or in the tide rips west of Grassholm. Identification is always a problem but species reliably reported include bottle-nosed whale, killer whale, pilot whale, Risso's dolphin, white-sided dolphin, bottle-nosed dolphin and common dolphin.

Bats: Dyfed is fortunate in its complement of bats; at least ten of Britain's 15 species are present. Some like the whiskered bat may be more numerous than in other parts of the country. Geology and land use may well have influenced the species composition and the present distribution of the bat population.

The limestone areas of the south have a small number of caves, which may have enabled the horseshoe bats to establish themselves. Both greater and lesser horseshoes were once common but are now less numerous. There is now thought to be only one colony of greater horseshoes for the whole of Dyfed and West Glamorgan. Two nursery roosts, in buildings, are used regularly and are protected by the NCC with the co-operation of the owners, and designated as Sites of Special Scientific Interest. Wogans Cavern in Pembroke Castle was probably the natural breeding centre with, in the 1950s, several thousand present, but it is no longer used. The Dyfed population is now probably no more than 400 strong. Lesser horseshoe bats are not so well known but a small number of roosts have been found in south Pembrokeshire. They seem to be absent from Carmarthenshire and Ceredigion, although there are unconfirmed reports for some of the northern lead mines.

Pipistrelles are widespread and may use as many as 80% of rural dwellings. Unlike the colonies in some other parts of the country where there are high densities in modern buildings, some may be present in the building throughout the year, the thick rubble filled walls of older properties providing suitable hibernation quarters. Both whiskered and Brandt's bats seem reasonably abundant, being found almost as often as pipistrelles in old buildings. Natterer's bat is less numerous than any of the previous species but is regularly encountered. Despite the abundance of water, Daubenton's bat (or Water Bat) is only locally distributed, preferring access to large areas of still water of either the few lakes or calmer stretches of the major rivers. The brown long-eared bat is perhaps the second commonest and is the only long-eared bat in the area.

The two other species recorded are the noctule and the barbastelle. The latter, a secretive bat, has only been found on two occasions. The large noisy bats flying in the summer evening skies have long been a familiar sight in Dyfed, but it was only in 1984 that a roost of the noctule bat was found, confirming its presence in the county. Since then two other roosts have been located in hollow trees. Colonies of noctules contain on average about 40 individuals, but 94 bats were counted out of one of these trees in 1985, while in addition a number of young bats did not emerge.

There are old reports of dead serotines being found and, with the recent confirmation for mid-Wales, it would seem safe to predict that they will soon be discovered for Dyfed. The county is well within the range of Leisler's bat, which should also eventually be found. Bechstein's bat, a nationally rare species, may also occur perhaps in some of the ancient and old woodlands of the area.

The landscape of Dyfed perhaps holds the key to its bat wealth. The grassland areas of the south-west fulfil part of the feeding needs of the greater horsehoe bats. Modern agriculture has impoverished some pastures, but areas such as the Castlemartin tank range are excellent and still produce a rich insect harvest for bats. The north of the county is more hilly, and small field systems often bounded by thick hedges may not lend themselves to improvement. Additionally copses and spinneys complement this environment, producing a combination of good sheltered feeding and roosting areas for bats.

Dragonflies: Recent studies have revealed that 25 species of dragonfly breed in Dyfed. Several of these have significant populations in West Wales, while one of the rarest British species, the southern damselfly, is present on several sites in Pembrokeshire and in particular occurs in large numbers on the wet heathlands around the Preseli Hills. A report by the Council of Europe also classified the beautiful demoiselle, small red damselfly and the club-tailed dragonfly as vulnerable on a European scale, and each of these have thriving populations.

Generally in Britain one would expect the species diversity of most insect groups to decrease northwards. This phenomena is reflected in odonata site assessment, where the number of species present on a site considered to be of regional importance decreases from 17 in southern England to seven in the Shetland Islands. Dyfed falls in the 12 species range. There are approximately 23 sites where the species present reach this number.

Our fauna is mainly of species that are considered to be of Mediterranean origin. Many of these are near their north-western limits in Dyfed, and we are fortunate to have populations of all three damselflies typical of this element: the scarce blue-tailed, southern and small red. The only communities thought to have originated from the arctic/boreal fauna are those of the upland lakes, bogs and pools of Carmarthenshire and Ceredigion, typified by the presence of the common hawker and black darter.

Different sites support different communities with each species having its own requirements. However, many sites are diverse enough to contain a number of different habitats, each of which supports a different species. The table shows in a generalised way the species that might be encountered at different types of site. Obviously, only good examples of each site type will support all the species listed. From the table it can be seen that some species occur over a wide range of sites while others are quite restricted. This, and the fact that some types of site are more widespread in Dyfed than others, reflects the number of individuals of each species present.

Common species include the banded demoiselle, large red, blue-tailed and azure damselflies, golden-ringed dragonfly and the common darter. Less common but widespread are the emerald damselfly, common and southern hawkers, emperor dragonfly, broad-bodied and four-spotted chasers. Less widespread but common where they occur are the banded demoiselle, southern damselfly, migrant hawker, keeled skimmer and black darter. Scarce species are the scarce blue-tailed, variable and small red damselflies, club-tailed and hairy dragonflies, black-tailed skimmer and the ruddy darter.

Site Type	Site example	
Running Water		
Large river - slow current	Afon Tywi	1,2,4,5,10,12,18,23
" - faster current	Afon Teifi	1,2,4,5,10,18,23
Small river - slow current	Afon Solva	1,2,4,5,18,23
" " - fast current	Eastern Cleddau	1,4,18
Soligenous mire	Brynberian	4,5,6,8,10,18,21,23
Very slow/almost stationary		
Vegetative deep mud substrate	Gors Fawr	4,5,6,10,14,18,21,23,25
Coastal dykes	Afon Leri	3,4,5,9,10,13,14,18,19,22,23,25
Still Water		
Coastal pools	Mere Pool duneslack	3,4,5,10,13,15,16,17,18,19,20,23,24
Lakes - lowland	Ludchurch	3,4,5,7,10,14,15,17,19,23
" highland	Llyn Llywernog	3,4,7,14,20,23,25
Marsh	Pentood	3,4,5,6,10,20,23
Heath pools and mires - Lowland	Dowrog	3,4,10,11,19,20,23,25
" " " " - Highland	Figyn Blaen Brefi	4,14,25

NOTE: Not all the species tabulated occur at the example site.

1	Beautiful demoiselle	9	Variable damselfly	17	Emperor dragonfly
2	Banded demoiselle	10	Azure damselfly	18	Golden-ringed dragonfly
3	Emerald damselfly	11	Small red damselfly	19	Broad-bodied chaser
4	Large red damselfly	12	Club-tailed dragonfly	20	Four-spotted chaser
5	Blue-tailed damselfly	13	Hairy dragonfly	21	Keeled skimmer
6	Scarce blue-tailed damselfly	14	Common hawker	22	Black-tailed skimmer
7	Common blue damselfly	15	Southern hawker	23	Common darter
8	Southern damselfly	16	Migrant hawker	24	Ruddy darter
				25	Black darter

Banded Demoiselle on Yellow Iris

ABOVE: Lanceolate spleenwort. (NCC) BELOW LEFT: Tenby daffodils. (PC) RIGHT: Bee orchid. (SBE)

ABOVE LEFT: Dune gentian in Great Britain, restricted to a few localities in Carmarthenshire, Pembrokeshire and Gower. (JWD) RIGHT: The field gentian is a species of unimproved pastures. (JWD) BELOW LEFT: Common spotted orchid. (JWD) RIGHT: Early purple orchid. (JWD)

The Conservation of West Wales

Red Kite over Cothi Valley

Wales, and in particular Dyfed, can boast of having one of the longest running, if not the longest single species bird protection scheme in the world. This concerns the red kite, formerly widespread in Great Britain until at least the early 19th century, but then so persecuted that it ceased to nest in either England or Scotland by 1870 and was much reduced in Wales. The late Colonel Morrey Salmon in his classic account of the red kite so succinctly described the reasons for the decline: 'Those who brought this about were the game preserving landowners and their indiscriminate and destructive agents, the gamekeepers, who pursued their calling with a callous indifference for anything but game and an appalling amount of cruelty exercised with the connivance and, in some cases, the active participation of their masters. As the species became rarer it was the target for skin and egg collectors . . . . . .'.

In 1893, the Professor of Botany at Aberystwyth, J.H. Salter, first became acquainted with the red kite and noted with alarm the annual nest robberies and the killing of adult birds. Ten years later he enlisted the help of the British Ornithologists' Union in an effort to save the small population from extinction and, as a result, the first kite committee was formed. Information on numbers present up to the late 1940s is at times a little vague, but probably the population never exceeded ten pairs, and in some periods may have been only half that number. With funds from private individuals and the Royal Society for the Protection of Birds (RSPB), bounties were paid to farmers and landowners on whose land kites nested successfully. The reward in the early years was five pounds, but was increased to £10 and eventually to £20.

In 1949 Captain and Mrs H.R.H. Vaughan, then resident in the upper Tywi Valley, persuaded the West Wales Field Society (WWFS) to resuscitate the Kite Committee with help from the RSPB, who in 1958 took full responsibility for the special measures and the finances involved. The new impetus of these post-war years paid dividends as the population slowly increased, so that by 1973 there were 26 pairs, the trend continuing to reach a peak of 47 pairs in 1982 when 23 young were reared, four less than in 1980 when 27 young were reared by 42 pairs. In 1985 40 pairs were present, of which 18 produced a total of 21 young. The pressures remain – not only habitat loss and disturbance, but also from egg collectors 'who pursue their illegal interests with a ruthlessness more characteristic of the worst excesses of the Victorian era'.

Except for the short-lived Pembrokeshire Natural History Society and the Dewisland Field Club during the first years of the century, and about which nothing much is now known, Dyfed was extremely slow in the formation of a natural history society when compared to most other parts of Great Britain. Perhaps because of the paucity of observers, the county fared rather badly when the Society for the Promotion of Nature Reserves (SPNR) published the first list of nationally important sites in 1915/16. Only two of these were in Dyfed; Kidwelly sand dunes (Tywyn Burrows) in Carmarthenshire was in Category B, while Tregaron Bog (Cors Caron) in Ceredigion was in Category C. Notable omissions included the Pembrokeshire islands.

In 1929 a similar listing gave Borth Bog (Cors Fochno) as the only site in Dyfed proposed as a nature reserve. At the same time the concept of National Parks was being debated, and the Pembrokeshire coast, together with that of Cornwall, were the two in the 'seaside parks' category. When the Dower list was published a few years later Pembrokeshire had been joined as a candidate for consideration by the slopes of Plynlimon.

On 26 February 1938 L.D. Whitehead of Ramsey and R.M. Lockley of Skokholm, with support from the RSPB, convened a public meeting in Haverfordwest to form the Pembrokeshire Bird Protection Society. One of the aims of the new Society was 'to completely stop the harmful activities of both professional and amateur collectors of eggs and young birds'. The *West Wales Guardian* printed a lengthy report of the formation of the Society with a heading 'Cliffs to be watched from Cardigan to Laugharne'. Members of the new society were urged to become watchers and, where special problems existed, a full-time person, paid up to 7s a day, would be employed. The instruction card for watchers makes interesting reading, and is a reminder of the fervour and enthusiasm surrounding the formation of the Society, which shortly afterwards launched an Appeal for £350 towards its work to include educational activities in schools.

Although the emphasis was on birds in Pembrokeshire, it was not long before far-sighted members were urging the new Society to include the neighbouring counties of Carmarthen and Ceredigion, and to develop along the lines of the Norfolk Naturalists' Trust, founded 12 years previously, by acquiring nature reserves. The outbreak of war less than two years later sadly curtailed these early initiatives, but even so the Society was able to announce in 1944 its first nature reserve, when a lease was taken of Cardigan Island: 'that piece of land surrounded by water' as it is quaintly described in the title deeds. To take account of its expanding role, the name was changed to the West Wales Field Society (WWFS) in 1945, and plans rapidly made to revive the Bird Observatory on Skokholm which R.M. Lockley had opened as the first in Great Britain in 1933.

Nationally, discussions were still taking place regarding a policy for nature reserves, Tregaron Bog (Cors Caron) having been joined on the list by Grassholm, Skokholm and Skomer. The Hobhouse Committee, in its proposals for National Parks, had suggested that the Pembrokeshire Coast be joined by the tidal headwaters of Milford Haven with its tree-lined creeks and quiet saltings, and Mynydd Preseli. The Brecon Beacons were proposed as another National Park, the western extremity of which lies within Carmarthenshire. The Pembrokeshire Coast National Park, including the two additional areas, was designated in 1949 and covers some 625 square kilometres with a coastline of 270 kilometres. The Brecon Beacons National Park, designated in 1957, includes 242 square kilometres in Carmarthenshire.

A major development was the acceptance by the government early in 1948 of the recommendations of the Huxley Committee for a Nature Conservancy Board, and in November of that year C. Diver became the first Director-General of the Nature Conservancy (NC). A Royal Charter was secured, which set out the responsibilities of the NC 'to provide scientific advice on the conservation and control of the natural flora and fauna of Great Britain; to establish, maintain, and manage nature reserves [these were to be known as national nature reserves (NNRs)] in Great Britain, including the maintenance of physical features of scientific interest; and to organise and develop the research and scientific services related thereto'.

In 1973 the NC was abolished and the Nature Conservancy Council (NCC) created, financed by and responsible to the Department of the Environment. In 1965 a Regional Office had been established for Dyfed and Powys in the grounds of the Welsh Plant Breeding Station at Plas Gogerddan just north of Aberystwyth. Previously its small staff had occupied cramped offices in

Swansea and Aberystwyth. Over the years the number has slowly grown, though by no means sufficiently to take account of rapidly increasing demands, particularly since the greater obligations, as a result of the Wildlife and Countryside Act with regard to Sites of Special Scientific Interest (SSSI). In 1986 there were, in addition to headquarters staff at Plas Gogerddan, reserve wardens, estate workers and assistant regional officers (AROs). The AROs are at the forefront of much of the NCC's work, and the help which they give to voluntary bodies cannot be too highly praised. At present there are four AROs in Dyfed, for the District Council areas of Ceredigion, Preseli and South Pembrokeshire, Carmarthen and Dinefwr, and Llanelli which is linked with Gower and the South Wales Region of NCC.

In 1946 Skokholm Bird Observatory re-opened, and for that year was run by the WWFS. Under an agreement with the owner, W. Sturt, a Field Centre was opened at the same time on Skomer. The charge for a week's full board and lodging on the islands was three guineas and the boat charge 5s. Visitors were reminded in the brochure to bring ration cards and 'their own preserves or the necessary coupons'. Although it was not possible to continue with the initiative on Skomer beyond 1946, much valuable experience had been gained and further pioneer work accomplished. The WWFS took the enterprising step of purchasing for £6,000 Dale Fort, the Victorian fort overlooking Dale Roads at the entrance to Milford Haven, and then leased it to the Council for the Promotion of Field Studies (CPFS), which appointed John Barrett as first warden in 1947. From that same year Skokholm was also run from Dale Fort, where each Saturday the boat was loaded with stores and visitors to cross to the island.

The 1950s saw a range of activities commenced by the WWFS; it even spread its wings into North Wales, and Merioneth remained linked with the south-west for nearly 20 years before in 1971 joining much more appropriately with the North Wales Naturalists' Trust (NWNT). Even in Merioneth the emphasis was still on seabirds and, as a result, the famous inland cormorant colony at Craig-yr-Aderyn or Bird Rock, near Tywyn was wardened. The WWFS, together with the West Midlands Bird Club, established the Bird Observatory on Bardsey Island off the tip of the Lleyn Peninsula in 1953.

A notable event by the WWFS in 1955 was the publication of a quarterly, later twice yearly journal *Nature in Wales*. First with its Welsh dragon cover, and later with a fine polecat, both drawn by Charles Tunnicliffe, the paper's reports and field notes have provided a wealth of material for members, and it has been the envy of many other natural history societies and trusts over the past 30 years. In 1963 the NWNT, and the Radnor section of the Hereford and Radnor Trust began to share the journal; hopes that the other Welsh trusts would join have unfortunately not been realised. The journal is now published by the National Museum of Wales.

In 1958 a further chapter of island involvement commenced when the WWFS purchased Skomer, the result of a special appeal which raised £4,000 towards the price. The island was resold to the NC, who in turn leased it to the WWFS at a nominal rental, built a house for the warden, and provided a tractor and ancillary equipment. The first resident warden took up his post in 1960 while in March 1986 the fifth to be appointed crossed Jack Sound for the first time. This continuity of employees over a quarter of a century is an indication of the attraction of living on Skomer for most of the year.

The year 1961 was an important one for the WWFS for, although it had acted like a trust from inception, all now agreed it was high time to recognise this with a change of name. At the Annual General Meeting in May that year the name West Wales Naturalists' Trust (WWNT) was adopted, with later the Manx shearwater as its emblem. Why, oh why, not the puffin – a bird with such instant appeal, recognised by everyone, and an ideal species with which to promote the Trust? Although again there was no change in role, there was one further change in name when in 1984 the Trust became the West Wales Trust for Nature Conservation (WWTNC), the name which appears on the title page of this book. This signifies to a much greater extent the work of the Trust, and also its firm links with the Royal Society for Nature Conservation, the national association of all 46 trusts in Great Britain.

In 1962 the WWNT commenced two important co-operative ventures. At Cors Fochno efforts were made to acquire parts of the bog, these being later sold or assigned to the NCC, who were eventually able to acquire the major part of the site. On the tidal reaches of the Western Cleddau the long process of establishing a regional Wildfowl Refuge was begun, with completion eventually effected in 1970, when the Home Secretary, then the Rt Hon James Callaghan PC MP, signed the Order under the Wild Birds Protection Act.

During this same period both the NCC and the RSPB were active. National Nature Reserves like those at Cors Caron and Cors Fochno were declared together with a National Wildfowl Refuge on the Dyfi. The RSPB, which had purchased Grassholm in 1947, established nature reserves at Dinas and Gwenffrwd in the upper Tywi Valley and at Ynyshir on the Dyfi.

National Nature Week in 1966 and European Conservation Year in 1970 brought new impetus to nature conservation in Wales. A liaison committee of representatives of all the Trusts was formed, later to be known as the Association of Trusts for Nature Conservation in Wales (ATNCW). Several roadside nature reserves were established by the WWNT with the co-operation of the Highways Authorities, although these are mere fractions of the total length of verge in the county. Few activities seem to arouse the protective concern of the general public more than the cutting of roadside verges, and there is always considerable disquiet expressed each year as the machines set forth, apparently mowing everything before them. Priority must be given to road safety. Not only this, all verges do need cutting; leave them, and brambles and coarse herbs will soon exclude many of the flowers. It is the trimming that is all-important. What would be ideal is a commitment from the Highway Authority to cut verges on a rotational system, first early, then late cuts in succeeding years, and so on.

An important development concerning the wider countryside took place in 1971 when the Carmarthen sections of the WWNT inaugurated a scheme of Farm Nature Reserves. Information on the natural history of farms was provided to the owner or occupier, together with advice as to how features of nature conservation interest could be maintained and enhanced. One difficulty soon encountered by a voluntary organisation with, at that time, no full time staff, was how to keep in touch with, and provide information for, the 50 or so Farm Nature Reserves which quickly entered the scheme, mostly in Carmarthenshire, with a few in Pembrokeshire. In addition there were other limitations; the words 'Nature Reserve' clearly put some farmers off asking for information and advice. These pioneer efforts were an important part of the progress towards the Farming, Forestry and Wildlife Advisory Groups (FFWAGs) which have now been established in all counties. In England many FFWAGs have full-time advisers, in Wales at present only one. There is much to be done; most farmers and landowners are anxious for assistance, an urgent need exists to expand this advisory service, and benefits to wildlife beyond all measure must surely be the result.

The period since the 1970s has seen a steady growth in all nature conservation activities in Dyfed. The WWNT now has eight sections responsible for local activities, while in the extreme south east the Llanelli Naturalists' Society (LNS) formed in 1973 carries out a good deal of Trust work, including the acquisition in 1978 of a small section of Ffrwd Farm Mire as a nature reserve. The Otter Haven Project, funded by the Vincent Wildlife Trust, appointed an officer for Wales who lives near Newcastle Emlyn. The Woodland Trust has acquired several sites in Dyfed but as yet has no local membership structure. The County Council and Llanelli Borough Council operate Country Parks which have a nature conservation interest. Both the Brecon Beacons National Park and the Pembrokeshire Coast National Park lease or own areas where nature conservation is the prime consideration. A greater emphasis on nature conservation matters by District Councils would be of considerable advantage.

Oil pollution has been a scourge of seabird colonies from the beginning of the century, with guillemot, razorbill and puffin the species most at risk. One of the first prosecutions under the Oil in Navigable Waters Act was as the result of evidence provided by R.M. Lockley in 1929 from

Skokholm, and this secured a £25 fine. The development during the 1960s of Milford Haven as one of Europe's major oil terminals heightened concern for the wellbeing of the Pembrokeshire seabird colonies. It is with some relief that a quarter of a century has elapsed without the coast of South Wales suffering disasters on the scale of the *Torrey Canyon* or *Amoco Cadiz*. Nevertheless there have been a number of incidents, those of the *Christos Bitas* in October 1978 and the *Bridgeness* of June 1985 being grim reminders of the damage that can occur when even small quantities of oil are spilled in a key area. The death toll from a mere 187 tons of oil lost by the *Bridgeness* was possibly as high as 5,000 adult and young seabirds.

The WWTNC has a system for covering beaches during an oil pollution incident and, at its centre at West Williamston, facilities for the cleaning and medication of oiled birds. This was established at Lower House Farm by the late Guy Hains and his wife Jean, with enthusiastic help from other members, in the wake of the *Christos Bitas* incident. Much of the construction work was carried out by a team from Pembrokeshire Community Industry and the Centre was opened by the Secretary of State for Wales, the Rt Hon Nicholas Edwards PC MP, on 10 April 1981. More recently the Centre received a Prince of Wales Award. In a major incident West Williamston would act as a holding unit for the onward transmission of oiled birds to the RSPCA; it can cope with smaller numbers and now has considerable expertise and experience in the handling and rehabilitation of oiled birds.

From no fulltime staff and no office in the early 1970s, the Trust now owns a substantial property in Haverfordwest which provides both office and an information and shop facility. There are now a Director, Administration Officer and Conservation Officer, together with a secretarial assistant. As the pressures on the wild places of Dyfed increase, the Trust needs all these staff if it is to respond to the challenge of the final decade of the 20th century. By then it will have passed its 50th birthday, its Golden Jubilee. The successes are for all to see, and a glance at the map, a look at Appendix I, will reveal the sites now protected and managed, but much more needs to be done to purchase and manage a complete range of the habitats which are found in West Wales. The disappointment must be that fewer than 5% of the population of the county are members. This must surely be the challenge, to increase membership of the Trust, for here is the key to both finance and influence in the years ahead. If you are not yet a member but have read this far you really ought to join. If you are already with us, there are many who are not, so please encourage them to support us.

Manx Shearwater

Red kite in upper Tywi valley (NM of W)

Appendix I: Nature Reserves in West Wales

Name and Description: The name by which the nature reserve is normally known may in some instances only be found on the larger scale maps. The description only provides brief habitat details. More information can be obtained from the management organisation, and in a few cases descriptive leaflets are available.

Map reference: From the 1:50,000 scale Ordnance Survey maps, the reference usually denotes the centre of the reserve and has no relevance to access points.

Area: Given in hectares, there are just under 2.5 acres to a hectare.

Manager: Although many nature reserves are owned by the bodies listed, others are leased, or held under management agreements with both private individuals and organisations like the Forestry Commission and National Trust.

BBNP	Brecon Beacons National Park, Glamorgan Street, Brecon, Powys LD3 7DP
LNS	Llanelli Naturalists' Society, Rhys-deg, Maesybont, Llanelli, Dyfed
NCC	Nature Conservancy Council, Dyfed Powys Regional Office, Plas Gogerddan, Aberystwyth, Dyfed SY23 3EB
PCNP	Pembrokeshire Coast National Park, County Offices, St Thomas's Green, Haverfordwest, Dyfed SA61 1QZ
RSPB	Royal Society for the Protection of Birds, Frolic Street, Newtown, Powys SY16 1AP
WT	Woodland Trust, Autumn Park, Dysart Road, Grantham, Lincolnshire NG31 6LL
WWTNC	West Wales Trust for Nature Conservation, 7 Market Street, Haverfordwest, Dyfed SA61 1NF

Status: A number of nature reserves are National Nature Reserves (NNR) designated by the Nature Conservancy Council, while others are Sites of Special Scientific Interest (SSSI). The Skomer Marine Nature Reserve at the time of publication was a voluntary MNR, but consultations are currently in progress towards eventual designation as a statutory site.

Access: Some nature reserves are open at all times to everyone, usually by means of designated footpaths, while one has a road intersecting it. In some cases only part of the nature reserve may be open. These are all indicated by the word 'Open' in the table. Nearly all the nature reserves managed by the WWTNC are open to members and to non-members who first obtain permission. Most of those managed by the other organisations are also available on application for a permit.

Name & Description	Map Ref.	Area (ha)	Manager	Status	Access
ALLT DOL LLAN Mixed woodland	SN420402	14	WWTNC		Open
ALLT CRUG GARN Fragments of old heath within coniferous plantation	SN518622	0.5	WWTNC	–	Open
ALLT FEDW CUTTING Railway cutting and embankment	SN665730	1	WWTNC	–	Open
ALLT RHYD-Y-GROES One of the finest examples of sessile oak woodland in southern central Wales	SN761482	62	NCC	NNR	–
BISHOP'S POND Ox-bow lake with interesting aquatic plants and invertebrates	SN445209	4	WWTNC	SSSI	–
BRUNT HILL Young oak/birch plantation in conifers	SM816074	3	WWTNC	–	Open
CAEAU LLETY CYBI Four small dry pastures, not ploughed for many years	SN603535	3.5	WWTNC	SSSI	Open
CARDIGAN ISLAND Island with gulls, small numbers of other seabirds and Soay sheep	SN160515	15	WWTNC	–	–
CASTLE WOODS Mixed woodland overlooking the Tywi watermeadows	SN625620	26	WWTNC	SSSI	Open
CEMAES HEAD Coastal heath and sea cliffs	SN135500	20	WWTNC	SSSI	Open
COED ALLT TROED-Y-RHIW FAWR Mixed woodland	SN41225	16	WT	–	–
COED CWM DDU Sessile oak woodland above the Afon Ceri	SN309429	26	WWTNC	–	Open
COED LLWYNGORAS Mixed woodland and riverbank	SN100390	25	WWTNC	–	Open
COED PENGLANOWEN Mixed woodland with an estate planting emphasis	SN610786	6	WWTNC	–	Open
COED PERTHNEIDR Mixed woodland	SN416584	3.5	WT	–	–
COED RHEIDOL An even aged sessile oak wood on either side of a steep gorge	SN745782	43	NCC	NNR	–
COED SIMDDE LWYD Sessile oak to the west, mixed woodland in the east	SN720786	28.5	WWTNC	SSSI	Open
COED TYDDYN DU The largest lowland broad-leaved woodland in Ceredigion	SN272426	19	WT	SSSI	–
COED Y CASTELL Ash, oak woodland	SN667193	16	BBNP	–	Open
COMINS CAPEL BETWS Acidic heath, unimproved grassland and marsh	SN615576	7	WWTNC	SSSI	–
CORS CARON Raised mire of botanical, ornithological and invertebrate interest	SN690640	768	NCC	NNR	Open
CORS GOCH Raised mire, alder carr and reedbed	SN372188	12	WWTNC	SSSI	–
CORS GORSGOCH Valley mire with large pond	SN482504	16	WWTNC	SSSI	–
CWM CLETTWR Sessile oak woodland in damp conditions above the Clettwr	SN670920	21	WWTNC	SSSI	Open
DINAS AND GWENFFRWD Wide range of communities including woodland and upland	SN788470	554	RSPB	SSSI	Open
DINEFWR DEER PARK Pasture woodland with rich lichen flora	SN613224	45	WWTNC	SSSI	–
DOWROG COMMON Wet and dry heath, marsh and open water	SM773268	83	WWTNC	SSSI	Open
FFRWD FARM MIRE Reed bed, marsh and unimproved pasture	SN420026	23	WWTNC LNS	SSSI	Open

Name & Description	Map Ref.	Area (ha)	Manager	Status	Access
FIGYN BLAEN BREFI Blanket mire at watershed of Afon Breifi	SN717547	45	WWTNC	SSSI	Open
GARNE TURNE ROCKS Old pasture with cromlech	SM979273	4	WWTNC	–	Open
GOODWICK MOOR Reedbeds, marsh, willow and alder carr	SM946377	18	WWTNC	–	Open
GRASSHOLM Remote island with gannet colony	SM599093	9	RSPB	SSSI	Open
GWAUN VALLEY WOODLANDS Mixed woodland	SN024339	50	PCNP	SSSI	Open
GWENFFRWD (See Dinas and Gwenffrwd)					
LLANNERCH ALDER CARR Alder carr	SN057353	2	WWTNC	SSSI	–
LLYN EIDDWEN Shallow lake, bog, upland grassland and mires	SN607674	55	WWTNC	SSSI	Open
LLYN FANOD Shallow lake similar in some respects to Llyn Eiddwen	SN604645	1	WWTNC	SSSI	Open
LLYN NANT-Y-BAI Small pools at disused lead mine	SN786447	1	WWTNC	–	Open
MARLOES MERE Rushy pasture with numerous pools	SM777082	11	WWTNC	SSSI	–
NANT MELIN Mixed broadleaved woodland and two small pastures	SN728467	3	WWTNC	SSSI	Open
OLD MILL GROUNDS Riverside marsh and woodland	SM953161	8	WWTNC	–	Open
OLD WARREN HILL Mixed woodland with Iron Age hill fort	SN613787	8	WWTNC	–	Open
PANT DA Young oak woodland on site of former larch plantation	SN670790	4	WWTNC	–	Open
PEMBROKE UPPER MILL POND Reed bed, willow carr and open water	SM993016	5	WWTNC	–	Open
PENDERI Sessile oak woodland on sea cliffs	SN550732	12.5	WWTNC	SSSI	Open
PENKELLY FOREST Superb ancient woodland, showing a range of habitats	SN132390	66	WWTNC	SSSI	Open
POOR MANS WOOD Sessile oak woodland	SN784356	18.5	WWTNC	–	Open
POPPIT REED BED Reed bed and willow carr	SN155484	6	PCNP	–	Open
PORTFIELD GATE QUARRY Two former quarry pools surrounded by scrub	SM923154	1	WWTNC	–	Open
RHOS FULLBROOK Herb-rich grassland with numerous flushes and marsh	SN668628	2.5	WWTNC	SSSI	–
RHOS GLYN-YR-HELYG Sedge-rich pasture, two basin mires and the Afon Grannell	SN498514	16	WWTNC	SSSI	–
RHOS LAWR CWRT Sedge rich pasture with numerous pingoes	SN411499	24	NCC	NNR	–
RHOS PIL-BACH Five damp meadows with well developed hedgerows and three ponds	SN367529	10.5	WWTNC	SSSI	Open
RHOS-Y-FFOREST Remnant of valley mire and flushes	SN618730	1	WWTNC	–	Open
ROSEMOOR Woodland, pasture, marsh and lake	SM873110	10	WWTNC	–	Open
SAM'S WOOD Estuarine cliff woodland	SN004095	12	WWTNC	–	Open

Name & Description	Map Ref.	Area (ha)	Manager	Status	Access
SKOKHOLM	SM735050	98	WWTNC	SSSI	–
Seabird island with superb maritime plant communities					
SKOMER	SM725095	294	WWTNC	NNR	Open
Seabird island, grey seals and maritime plant communities					
SKOMER MARINE RESERVE	SM752090	1478	–	MNR	Open
A wide range of sub-littoral habitats					
STACKPOLE	SR985945	797	NCC	NNR	Open
Lakes, sea cliffs and dunes					
St. MARGARET'S	SS120974	7	WWTNC	SSSI	–
Largest cormorant colony in Wales					
TEIFI FORESHORE	SN190453	14	WWTNC	SSSI	Open
Tidal foreshore and reed fringed lagoon					
TREGYB WOODLAND	SN641217	27	WT	SSSI	–
One of the largest blocks of broadleaved woodland in south-east Dyfed					
TYCANOL	SN091370	74	NCC	NNR	–
Woodland with glades, small fields and moorland with rocky tors					
TY NEWYDD	SN753335	0.25	WWTNC	–	Open
Roadside verge					
WERN DDU WOOD	SN373179	1	WWTNC	–	Open
Mixed woodland					
WEST HOOK CLIFFS	SM762092	8	WWTNC	–	Open
Coastal heath and scrub					
WEST WILLIAMSTON	SN030060	19	WWTNC	SSSI	Open
Salt marsh, limestone scrub and grassland and woodland					
WESTERN CLEDDAU MIRE	SM894307	12	WWTNC	SSSI	–
Flood plain mire, marsh and carr	SM907317	26	NCC	NNR	–
YNYSHIR	SN683963	255	RSPB	SSSI	–
Woodland, moorland, saltmarsh and estuary					
Y GOYALLT	SN781442	6	WWTNC	–	Open
Sessile oak woodland					

Appendix II: References and Further Reading

The Journal *Nature in Wales*, published since 1955, contains numerous articles and notes on the wildlife of West Wales, too many to list here. It should be consulted by anyone wishing to learn further about the county. Although complete runs of *Nature in Wales* are scarce nevertheless many numbers are still available for sale from the WWTNC and it should have a place on every naturalist's bookshelves. More recently the excellent duplicated newsletters of the Llanelli Naturalists Society include a wealth of information, mainly concerning south-east Carmarthenshire, and are a model for any society. Individual papers have not been listed. Of immense importance is *A Natural History Bibliography of Pembrokeshire* by John Comont, published in 1980. Although not available for sale, copies are to be found in the main libraries of Dyfed. Similar Bibliographies are long overdue for Carmarthenshire and Ceredigion.

The Pembrokeshire islands of Skokholm and Skomer have a wealth of literature. Annual reports were published for the former up to 1976, with Skomer material included from 1973. More recently, the Bulletin of the Friends of Skokholm and Skomer includes papers, notes, and the island bird reports.

BORROW, G. (1970) Wild Wales.

BOWEN, E.G. (1965) Wales: A Physical, Historical and Regional Geography. Methuen.

BOWEN, E.G., CARTER, H., TAYLOR, J.A. (1968) Geography at Aberystwyth. University of Wales Press.

BUCHANAN, J., & FULLER, M., Pembrokeshire Ancient Woodlands Survey 1978. Report by Nature Conservancy Council and West Wales Naturalists Trust.

BUXTON, J. & LOCKLEY, R.M. (1950) Island of Skomer. Staples.

COKER, S. & FOX, A. (1986) Dragonflies of Dyfed. Clarbeston Road.

COMONT, J.C. (1980) A Natural History Bibliography of Wales. Pembrokeshire Biological Records Centre.

CONDRY, W.M. (1981) The Natural History of Wales. Collins.

DAVIS, P.E. & RODERICK, H. (1985) Ceredigion Bird Report 1982-1983. Haverfordwest. (Includes details of the status of all birds recorded in Ceredigion and so up-dates Ingram, G.S. et al. 1966).

DAVIS, T.A.W. (1970) Plants of Pembrokeshire. Haverfordwest.

DAVIS, T.A.W. Botanical Diaries 1962-1980. Two volumes held by Botany Department, National Museum of Wales.

DONOVAN, J.W. & REES, G.H. (1986) Pembrokeshire Bird Report 1985. Haverfordwest.

ELLIS, R.G. (1983) Flowering Plants of Wales. Cardiff.

FALCONER, R.W. (1848) Contribution towards a Catalogue of Plants Indigenous to the Neighbourhood of Tenby. Longmans.

FENTON, R. (1811) A Historical Tour through Pembrokeshire.

GEORGE, T.N. (1970) British Regional Geology, South Wales. HMSO.

GERALD OF WALES (1978) The Journey through Wales and the Description of Wales. Penguin Books.

GOSSE, P.H. (1856) Tenby – A seaside holiday. Van Voorst.

HISCOCK, K. (1981) South-west Britain Sublittoral Survey. Orielton.

HOWELLS, ROSCOE (1968) The Sounds Between. Llandysul.
HYDE, H.A., WADE, A.E. & HARRISON (5th 1969) Welsh Ferns, Clubmosses, Quillworts and Horsetails. Cardiff.
HYWEL-DAVIES, J. & THOM, V. (1984) The Macmillan Guide to Britain's Nature Reserves. London.
INGRAM, G.C.S., SALMON, H. MORREY (1954) A Hand List of the Birds of Carmarthenshire. Cardiff.
INGRAM, G.C.S., SALMON, H. MORREY & CONDRY, W.M. (1966) The Birds of Cardiganshire. Haverfordwest.
JERMY, A.C., ARNOLD, H.R., FARRELL, L., PERRING, F.H. (1978) Atlas of the Ferns of the British Isles. London.
JOHN, B. (1976) Pembrokeshire. David and Charles.
LACEY, W.S. Ed (1970) Welsh Wildlife in Trust. North Wales Naturalist's Trust.
LEWIS, C.A. Ed. (1970) The Glaciation of Wales and Adjoining Region. Longman.
LEWIS, W.J. (1969) Cardiganshire Historical Atlas. Aberystwyth.
LINNARD, W. (1982) Welsh Woods and Forests. Cardiff.
LLOYD, B. Pembrokeshire Journals 1925-1939. Two volumes held by Botany Department, National Museum of Wales.
LOCKLEY, R.M., INGRAM, C.S. & SALMON, H. MORREY (1949) The Birds of Pembrokeshire. Haverfordwest.
LOCKLEY, R.M (1969) Pembrokeshire. Robert Hale & Co.
LOCKLEY, R.M. (1970) The Naturalist in Wales. David and Charles.
LOCKLEY, R.M. (1981) The Island. Penguin Books.
MATHESON, C. (1932) Changes in the Fauna of Wales Within Historic Times.
MAY, R.F. (1967) A List of the Flowering Plants and Ferns of Carmarthenshire. Haverfordwest.
NELSON-SMITH, A. & CASE, R.M. (1984) Aspects of the Shore and Sublittoral Ecology of the Daucleddau Estuary, Milford Haven. Zoological Journ. of Linnean Society 80: 177-190.
OWEN, GEORGE (1603) The Description of Pembrokeshire by George Owen of Henllys. Edited by Henry Owen. 1906. Clark.
PERRING, F.H. & WALTERS, S.M. (1976) Atlas of the British Flora. London.
PERRING, F.H. & FARRELL, L (1977) British Red Data Book: 1 Vascular Plants. Lincoln.
POWELL, H.T. et al. (1979) Report on the Shores of South West Wales. A report to the Nature Conservancy Council by the Marine Biological Association.
REES, W. (1966) An Historical Atlas of Wales from Early to Modern Times.
ROBERTS, D.H.V. (1986) The Carmarthenshire Bird Report 1985. Haverfordwest.
STAMP, L.D. (1946) The Land of Britain and How it is Used. Longmans Green & Co.
WILLIAMS, M. (1977) The South Wales Landscape. Hodder & Stoughton.
WILLIAMS, M. (1980) National Atlas of Wales. University of Wales.

Key to Caption Credits

The following abbreviations are used in the caption credits. The West Wales Trust for Nature Conservation is indebted to all the artists and photographs who have made their work available and which is reproduced here with their permission.

MA	Michael Alexander
SB	Sian Boisevain
JB	Jean Buchanan
BBNP	Brecon Beacons National Park
PC	Peter Chapman
HTC	Steve Conway
JC	John Comont
PC	Peter Chapman
U of C	University of Cambridge
JWD	Jack Donovan
MOD	Ministry of Defence
SBE	Stephen Evans
RE	Ron Elliot
DF	Dawn Fisher
LG	Liz Gardner
JH	John Harvey
RH	Roscoe Howells
SJ	Studio Jon
SL	Sue Linney
NCC	Nature Conservancy Council
NMW	National Museum of Wales
HP	Harold Platt
RP	Richard Pryce
SHR	Stewart Roberts
JPS	John Savidge
BS	Brian Southern
DRS	David Saunders
JT	Jeffrey Taylor

Index

Species

Scientific names have been excluded from the text but are included in the index as are Welsh names. In the case of birds I have followed *Rhester o Adar Cymru* by P. Hope Jones and E.V. Breeze Jones. For plants the *Flowering Plants of Wales* by R.G. Ellis and *Welsh Ferns* by H.A. Hyde, A.E. Wade and S.G. Harrison have been my guide.

1. VERTEBRATES OTHER THAN BIRDS

2. BIRDS

Species	*Welsh name*	*Scientific name*	*Page refs.*
Bittern	Aderyn y Bwn	*Botaurus stellaris*	129
Blackcap	Telor Penddu	*Sylvia atricapilla*	34,105
Bluethroat	Bronlas	*Luscinia svecica*	130
Bunting, Black-headed	Bras Penddu	*Emberiza melanocephala*	114
Cirl	Bras Ffrainc	*Emberiza cirlus*	129,130
Corn	Bras zr Yd	*Emberiza calandra*	129,130
Lapland	Bras y Gogledd	*Calcarius lapponicus*	130
Little	Bras Lleiaf	*Emberiza pusilla*	130
Ortolan	Bras y Gerddi	*Emberiza hortulana*	130
Reed	Bras y Cyrs	*Emberiza schoeniclus*	25,85
Rustic	Bras Gwledig	*Emberiza rustica*	114
Snow	Bras yr Eira	*Plectrophenax nivalis*	109,130,131
Buzzard	Bwncath	*Buteo buteo*	24,34,37,82, 91,129,131
Chaffinch	Ji-binc	*Fringilla coelebs*	25
Chiffchaff	Siff-saff	*Phylloscopus collybita*	115
Chough	Bran Goesgoch	*Pyrrhocorax pyrrhocorax*	*3*,47,*89*,91,92, *103*,110,111
Cormorant	Mulfran	*Phalacrocorax carbo*	92,110,111,129
Corncrake	Rhegen yr Yd	*Crex crex*	50,56,129, 130,131
Courser, Cream-coloured	Rhedwr y Twxni	*Cursorius cursor*	129
Crake, Spotted	Rhegen Fraith	*Porzana porzana*	114
Crossbill	Gylfin Groes	*Loxia curvirostra*	128,129,130
Cuckoo	Cog	*Cuculus canorus*	25,131
Curlew	Gylfinir	*Numenius arquata*	25,*28*,50,*55*,85, 96,118,129,131
Diver, Red-throated	Trochydd Gyddfgoch	*Gavia stellata*	129
Dipper	Bronwen y Dwr	*Cinclus cinclus*	25,131
Dove, Collared	Turtur Dorchog	*Streptopelia decaocta*	105,128,130
Turtle	Turtur	*Streptopelia turtur*	115,129
Duck, Ruddy	Hwyaden Goch	*Oxyura jamaicensis*	84,129
Tufted	Hwyaden Gopog	*Aythya fuligula*	84,129,130,131
Dunlin	Pibydd y Mawn	*Calidris alpina*	25,85,96,129
Egret, Little	Creyr Bach	*Egretta garzetta*	131
Eider	Hwyaden Fwythblu	*Somateria mollissima*	85
Fieldfare	Socan Eira	*Turdus pilaris*	131
Firecrest	Dryw Penfflamgoch	*Regulus ignicapillus*	130
Flycatcher, Pied	Gwybedog Brith	*Ficedula hypoleuca*	34,*35*,37,82, 115,129,130, 131
Red-breasted	Gwybedog Brongoch	*Ficedula parva*	109,130
Spotted	Gwybedog Mannog	*Muscicapa striata*	115
Fulmar	Aderyn-Drycin y Graig	*Fulmarus glacialis*	107,110,128, 129,130
Gadwall	Hwyaden Lwyd	*Anas strepera*	129
Gannet	Hugan	*Sula bassana*	107,108,109, *113*,*123*
Garganey	Hwyaden Addfain	*Anas querquedula*	56
Godwit, Bar-tailed	Rhostog Gynffonfrith	*Limosa lapponica*	85
Black-tailed	Rhostog Gynffonddu	*Limosa limosa*	85
Goldeneye	Hwyaden Lygad-aur	*Bucephala clangula*	131
Goosander	Hwyaden Ddanheddog	*Mergus merganser*	*75*,128,130
Goose, Canada	Gwydd Canada	*Branta canadensis*	128,130
Greylag	Gwydd Wyllt	*Anser anser*	128,129
White-fronted	Gwydd Dalcen-wen	*Anser albifrons*	25,56,72,84, 129,130
Goshawk	Gwalch Marth	*Accipiter gentilis*	128,129,130, 131
Grebe, Great Crested	Gwyach Fawr Gopog	*Podiceps cristatus*	74,84,128,129, 130
Greenfinch	Llinos Werdd	*Carduelis chloris*	105
Grosbeak, Rose-breasted	Gylfindew Brongoch	*Pheucticus ludovicianus*	114,130
Grouse, Black	Grugiar Ddu	*Lyrurus tetrix*	25,*29*,56,129, 130
Red	Grugiar	*Lagopus lagopus*	25,129,130
Guillemot	Gwylog	*Uria aalge*	92,109,110, 112,115,130, 144
Guillemot, Black	Gwylog Ddu	*Cepphus grylle*	130
Gulls, Black-headed	Gwylan	*Larus ridibundus*	25,*45*,118,129
Glaucous	Gwylon y Gogledd	*Larus hyperboreus*	84
Great Black-backed	Gwylan Gefnddu Fwyaf	*Larus marinus*	107,109,112
Herring	Gwylan y Penwaig	*Larus argentatus*	*45*,106,109, 118,129,130
Iceland	Gwylan yr Arctig	*Larus glaucoides*	131
Lesser Black-backed	Gwylan Gefnddu Leiaf	*Larus fuscus*	118,130,131
Little	Gwylan Fechan	*Larus minutus*	130
Mediterranean	Gwylan Mor y Canoldir	*Larus melanocephalus*	131
Sabine's	Gwylan Sabine	*Larus sabini*	131
Harrier, Hen	Bod Tinwen	*Circus cyaneus*	24,61,84,131
Marsh	Bod y Gwerni	*Circus aeruginosus*	129
Montagu's	Bod Montagu	*Circus pygargus*	56,61
Heron, Purple	Creyr Porffor	*Ardea purpurea*	56
Hobby	Hebog yr Ehedydd	*Falco subbuteo*	130
Hoopoe	Copog	*Upupa epops*	114
Kestrel	Cudyll Coch	*Falco tinnunculus*	24,131
Kite, Red	Barcud	*Milvus milvus*	24,37,129,130, *141*,141,*150*
Kittiwake	Gwylan Goesddu	*Rissa tridactyla*	92,*opp 97*,109, 110,112,115, *124*,130
Knot	Pibydd yr Aber	*Calidris canutus*	85
Lapwing	Cornchwiglen	*Vanellus vanellus*	25,50,*53*,*87*, 111,129,130
Lark, Short-toed	Ehedydd Llwyd	*Calandrella cinerea*	130,131
Linnet	Llinos	*Acanthis cannabina*	25,48,94
Magpie	Pioden	*Pica pica*	115
Mallard	Hwyaden Wyllt	*Anas platyrhynchos*	96,129,131
Martin, House	Gwennol y Bondo	*Delichon urbica*	91
Sand	Gwennol y Glennydd	*Riparia riparia*	115,130
Merganser, Red-breasted	Hwyaden Frongoch	*Mergus serrator*	85,129,130
Merlin	Cudyll Bach	*Falco columbarius*	*21*,24,61,84, 129,130,131
Nightingale	Eos	*Luscinia megarhynchos*	129
Nightjar	Troellwr Mawr	*Caprimulgus europaeus*	129,130,131
Nuthatch	Delor y Cnau	*Sitta europaea*	115,130,131
Oriole, Baltimore	Euryn Baltimore	*Icterus galbula*	130
Golden	Euryn	*Oriolus oriolus*	114
Osprey	Gwalch y Pysgod	*Pandion haliaetus*	84
Ouzel, Ring	Mwyalchen y Mynydd	*Turdus torquatus*	25,115,129,131
Owl, Barn	Tylluan Wen	*Tyto alba*	47,129,130
Little	Tylluan Fach	*Athene noctua*	130
Long-eared	Tylluan Gorniog	*Asio otus*	129
Scops	Tylluan Scops	*Otus scops*	114
Short-eared	Tylluan Glustiog	*Asio flammeus*	61,84,109,118, 129,131
Oystercatcher	Pioden y Mor	*Haematopus ostralegus*	85,95,96,*opp* 97,103,107,129
Partridge, Grey	Petrisen Llwyd	*Perdix perdix*	50,128,130,131
Red-legged	Petrisen Goesgoch	*Alectoris rufa*	128
Peregrine	Hebog Tranmor	*Falco peregrinus*	24,84,91,*103*, 129,130,131
Petrel, Storm	Pedryn Drycin	*Hydrobates pelagicus*	130
Pheasant	Ffesant	*Phasianus colchicus*	*47*,*51*
Pintail	Hwyaden Lostfain	*Anas acuta*	85,109,130
Pipit, Meadow	Corhedydd y Waun	*Anthus pratensis*	25,107,131
Olive-backed		*Anthus hodgsoni*	114
Red-throated	Corhedydd Gyddfgoch	*Anthus cervinus*	114
Richard's	Corhedydd Richard	*Anthus novaeseelandiae*	130
Rock	Corhedydd y Graig	*Anthus spinoletta*	107
Tawny	Corhedydd Melyn	*Anthus campestris*	130
Tree	Corhedydd y Coed	*Anthus trivialis*	25,37,82,129
Plover, Golden	Cwtiad Aur	*Pluvialis apricaria*	25,*28*,129,130
Kentish	Cwtiad Caint	*Charadrius alexandrinus*	130
Killdeer		*Charadrius vociferus*	129,*133*
Lesser Golden		*Pluvialis dominica*	114
Little Ringed	Cwtiad Torchog Bach	*Charadrius dubius*	129,130
Ringed	Cwtiad Torchog	*Charadrius hiaticula*	85
Pochard	Hwyaden Bengoch	*Aythya ferina*	84,129,130,131
Puffin	Pal	*Fratercula arctica*	*3*,*opp 5*,*8*,107, 108,110,113, 116,117,118, 130,143,144
Quail	Sofliar	*Coturnis coturnix*	56
Raven	Cigfran	*Corvus corax*	24,*29*,91,107, 129,131
Razorbill	Llurs	*Alca torda*	92,*opp 97*,107, 109,110,112, 130,144
Redpoll	Llinos Bengoch	*Acanthis flammea*	129,131
Redshank	Pibydd Coesgoch	*Tringa totanus*	25,85,96,129, 131
Redstart	Tingoch	*Phoenicurus phoenicurus*	25,34,37,115, 129,131
Redstart, Black	Tingoch Du	*Phoenicurus ochruros*	109,115
Redwing	Coch dan-aden	*Turdus iliacus*	131
Robin	Robin Goch	*Erithacus rubecula*	115
Rook	Ydfan	*Corvus frugilegus*	25,109,115
Rosefinch, Scarlet	Llinos Goch	*Carpodacus erythrinus*	114

3. PLANTS VASCULAR AND NON VASCULAR

Sedge, Bottle	Hesgen Chwysigen Naidd Ylfain	*Carex rostrata*	74
Dioecious	Hesgen Ysgar	*Carex dioica*	128
Fibrous Tussock	Hesgen Manedafedd	*Carex appropinquata*	*Opp 64*,128
Flat-sedge	Hesgen Wastad	*Blysmus compressus*	96
Great Fen	Llemfrwynen	*Cladium mariscus*	56,60,128
Greater Pond	Hesgen Braff-dywysennog	*Carex riparia*	82
Greater Tussock	Hesgen Rafunog Fwyaf	*Carex paniculata*	128
Lesser Pond	Hesgen Ganiolig-dywysennog	*Carex acutiformis*	128
Long-stalked Yellow	Hesgen Felen Paladr Hir	*Carex lepidocarpa*	59
Mud	Hesgen Eurwerdd	*Carex limosa*	23
Slender	Hesgen Feindwf	*Carex lasiocarpa*	59,128
Soft-leaved	Hesgen Mynydd-dir	*Carex montana*	24
Spring	Hesgen Gynnar	*Carex caryophylla*	81
Tufted	Hesgen Oleulas Sythddail	*Carex elata*	128
Water	Hesgen y Dŵr Mynyddog	*Carex aquatilis*	74,125
White-beak	Corsfrwynen Wen	*Rhynchospora alba*	23,56,81
Service Tree, Wild	Criafolen Wyllt	*Sorbus torminalis*	36
Shoreweed	Beistonnell	*Littorella uniflora*	23,29,114
Silverweed	Tinllwyd	*Potentilla anserina*	115
Skullcap, Lesser	Cycyllog Bach	*Scutellaria minor*	82,114
Soloman's Seal	Sêl Soloman	*Polygonatum multiflorum*	128
Sorrel, Sheep	Suran yr Yd	*Rumex acetosella*	114
Spearwort, Greater	Llafnlys Mawr	*Ranunculus lingua*	60,125,128
Lesser	Llafnlys Bach	*Ranunculus flammula*	114
Speedwell, Spiked		*Veronica spicata*	92,125
Sphagnum	Migwyn	*Sphagnum species*	23,*Opp 48*,55,*64*
Spleenwort, Forked	Duegredynen Fforchlog	*Asplenium septentrionale*	23,127
Green	Duegredynen Werdd	*Asplenium viride*	23,81
Lanceolate	Duegredynen Llafnbig	*Asplenium billotii*	*139*
Spindle	Piswydden	*Euonymus europaeus*	*Opp 48*,81
Spruce, Sitka	Spriwsen Sitka	*Picea sitchensis*	22,131
Spurge, Dwarf	Fflamgoed Eiddil Flaenfain	*Euphorbia exigua*	126
Spurrey, Rock Sea	Tywodwlydd y Morgreigau	*Spergularia rupicola*	*Opp 96*

Squill, Spring	Seren y Gwanwyn	*Scilla verna*	92,*93*,101,126
Starwort, Water	Brigwydd	*Callitriche species*	70
St John's Wort, Marsh	Eurinllys y Gors	*Hypericum elodes*	82,114
Wavy	Eurinllys Tonnog-ddail	*Hypericum undulatum*	56,*57*
Stonewort			*Opp 64*,74,81,126
Sundew, Long-leaved	Gwlithlys Mawr	*Drosera anglica*	128
Oblong-leaved	Gwlithlys Mirddail	*Drosera intermedia*	*Opp 48*
Round-leaved	Gwlithlys	*Drosera rotundifolia*	23,*57*,128
Sycamore	Masarnen	*Acer pseudoplatanus*	47
Thistle, Carline	Ysgallen Siarl	*Carlina vulgaris*	81
Meadow	Ysgallen Mignwern	*Cirsium disectum*	82,125
Thrift	Clustog Mair	*Armeria maritima*	92,96,127
Thyme, Wild	Teim Gwyllt	*Thymus praeox*	81,95
Toadflax, Prostrate	Gingroen Ymlusgol	*Linaria supina*	85,*87*,125
Toothwort	Deintlys Cennog	*Lathraea squamaria*	81,125
Tormentil	Tresgl y Moch	*Potentilla erecta*	82
Traveller's Joy	Cudd y Coed	*Clematis vitalba*	81
Trefoil, Hairy Bird's-foot	Pysen y Ceirw Flewog	*Lotus subbuliforus*	127
Valerian, Marsh	Triaglod y Gors	*Valerianella dioica*	82,125
Vetch, Kidney	Plucen Felen	*Anthyllis vulneraria*	82,92
Vetch, Wood Bitter	Pysen y Coed	*Vicia orobus*	126
Violet, Early Dog	Gwiolydd y Goedwig	*Viola reichenbachiana*	81,82
Pale Heath	Millyn Welw Grugog	*Viola lactea*	127
Water-crowfoot, Three-lobed	Egyllt y Llynnoedd	*Ranunculus tripartitus*	70,127
Water Lilly, White	Lili-ddŵr Wen	*Nymphaea alba*	74,81
Yellow	Lili-ddŵr Felen	*Nuphar lutea*	81
Water Plantain	Dŵr-lyriad	*Alisma plantago-aquatica*	82
Floating	Dŵr-lyriad Nofiadwy	*Luronium natans*	60
Whin, Petty	Cracheithin	*Genista anglica*	56,82,125
Whitlow Grass, Common	Llys y Bystwn	*Erophila verna*	81
Willow, Creeping	Corhelygen	*Salix repens*	84
Dwarf	Helygen Leiaf	*Salix herbacea*	22,125
Woodruff	Briwydden Bêr	*Galium odoratum*	34,81

4. INVERTEBRATES

Species	*Welsh name*	*Scientific name*	*Page refs.*
Argus, Brown	Gwrmyn Glas	*Aricia agestis*	84,94
Beetle, Bloody-nosed		*Timarcha tenebricosa*	94
Oil		*Meloe proscarabaeus*	94
Strandline		*Eurynebria Complanata*	84
Wasp		*Clytus arietus*	81
Beer Barrel Shell		*Acteon tornatilis*	95
Blue, Common	Glesyn Cyffredin	*Polyommatus icarus*	82
Small	Glesyn Bach	*Cupido minimus*	81,84,94
Brimstone		*Gonetteryx rhamni*	*Opp 112*
Bush Cricket, Great		*Tettigonia viridissima*	59
Chafer, Bee		*Trichius fasiatus*	81
Rose		*Cetonia aurata*	94
Chaser, Broad-bodied	Picellwr Boliog	*Libellula depressa*	*Opp 113*
Four-spotted	Picellwr Pedwar Nod	*Libellula quadrimaculata*	138
Cone-head, Short-winged		*Conocephalus dorsalis*	85
Copper, Small	Copor Bach	*Lycaena phlaeas*	82,*Opp 112*
Damselfly, Azure	Coenagrion Gyffredin	*Coenagrion puella*	138
Blue-tailed	Ischnura Cyffredin	*Ischnura elegans*	138
Common Blue	Mursen Las Gyffredin	*Enallagma cyathigerum*	138
Emerald	Lestes Werdd	*Lestes sponsa*	138
Scarce Blue-tailed	Ischnura Brin	*Ischnura pumilio*	138
Small Red	Mursen Fach Goch	*Ceriagrion tenellum*	*Opp 113*,138
Southern	Coenagrion Benfro	*Coenagrion mercuriale*	138
Variable	Coenagrion Amrywiol	*Coenagrion puella*	*Opp 113*
Large Red	Mursen Fawr Goch	*Pyrrhosoma nymphula*	138
Darter, Black	Picellwr Du	*Sympetrum danae*	*Opp 113*,81,138
Common	Picellwr Cyffredin	*Sympetrum striolatum*	*Opp 113*,138
Ruddy	Picellwr Rhudd	*Sympetrum sanguineum*	138

Demoiselle, Banded	Agrion Wych	*Calopteryx splendens*	*138*,138
Beautiful	Agrion Dywyll	*Calopteryx virgo*	*Opp 113*,138
Dragonfly, Club-tailed	Gwesyn Cnwpgwt	*Gomphus vulgatissimus*	72,138
Emperor	Yr Ymerawdwr	*Anax imperator*	138
Golden-ringed	Gwesyn Eurdorch	*Cordulegaster boltonii*	*Opp 113*
Hairy	Gwesyn Blewog	*Brochytron pratense*	85
Eurynebria		*Eurynebria complanata*	95
Fritillary, Dark Green	Britheg Werdd	*Argynnis aglaja*	84,94
Marsh	Britheg y Gors	*Euphydryas aurinia*	49,82,84, Opp 112
Pearl-bordered	Britheg Beriog	*Bolaria euphrosyne*	34
Silver-washed	Brotheg Arian	*Argynnis paphia*	34,84
Grayling	Iâr Fach y Graig	*Hipparchia semele*	84,94,*Opp 112*
Hairstreak, Brown	Brithribin Frown	*Thecla betulae*	48,*Opp 112*
Green	Brithribin Werdd	*Callophrys rubi*	94
White-letter	Brithribin Wen	*Strymonidia w-album*	34,*37*
Hawker, Common	Gwasneidr Glas	*Aeshna juncea*	138
Migrant	Gwasneidr Tramor	*Aeshna mixta*	138
Southern	Gwasneidr Llachar	*Aeshna cyanea*	138
Heath, Large	Gweunlöyn Mawr	*Coenonympha tullia*	56,*Opp 112*
Moth, Rosy Marsh	Gwyfyn Rhuddog y Gors	*Eugraphe subrosea*	56
Pillbug, White		*Armadillidium album*	84
Shell, Razor	Cragen Blisgyn	*Pharus legumen*	95
Scallop, Great	Cragen Fylchog	*Pecten maximus*	118
Sea Fan		*Pennatula phosphorea*	118
Sea Fingers, Red	Cwrel Coch	*Alycyonium glomeratum*	118
Skimmer, Keeled	Orthetrum Rhesog	*Orthetrum coerulescens*	*Opp 113*
Skipper, Small	Gwibiwr Bach	*Thymelicus sylvestris*	82
White, Marbled	Iar Fach Gleisiog	*Melanargia galathea*	81,82,*83*,84,95

General Index

Note: Plant and animal species are indexed in the species index. References to illustrations are in italics.

Subscribers

Presentation Copies

1 The West Wales Trust for Nature Conservation
2 Royal Society for Nature Conservation
3 The Nature Conservancy Council
4 Brecon Beacons National Park
5 Dyfed County Council
6 Ronald Lockley

7 David Saunders
8 Clive & Carolyn Birch
9 Liz Gardner
10 Joe Lewis
11 Anne Burn
12 Peter Davis
13 Bob Phillips
14 Stephen Coker
15 Jack Donovan
16 Richard Pryce
17 Stephen Evans
18 Miss Dorothy Phillips
19 Carmarthen Museum
20 Mrs A. Simmonds
21 The Rev Canon T. Hughie Jones
22–25 Llyfrgell Dyfed—Ceredigion
26 L.R.H. Griffiths
27 K.M. Jones
28 D.S. Hughes
29 Gavin Hall
30–31 Dr Stephanie J. Tyler
32 Mrs P.J. Parsons
33 Lyn Lewis Dafis
34 Cedric G. Conolly
35 K.M. Pollard
36 Dillwyn Miles
37 Paul & Rosalind Beck
38 Michael & Anne Brown
39 Victoria & Albert Museum
40 London Guildhall Library
41 Ivan Pedley
42 Mrs M. Jones
43 R. & J. Thomas
44 Ivan G. Williams
45 Alan J.M. Grieve
46–47 Michael Wells
48 Sue Dalton
49 Ian Campbell
50 I.K. Morgan
51 Miss Margaret Batt
52 M.E. Tiller
53 W.H. Gravell
54–55 Richard D. Pryce
56 D.G. Melton
57 R.E. Saunders
58 Mrs M.D. Ravenswood
59 Joseph Lewis
60 A.P.R. East
61 C. Fuller
62 Fiona Benn
63 Sheridan Newton
64 Mr & Mrs W.H. Cureton
65 Robert Reeves
66 Mr Finbow
67 D.J. Roberts
68 Dr & Mrs W.G.G. Loyn
69 I.W.R. Spriggs
70 Mrs D.V. Newman
71 Mrs C. Collyer
72 Eileen Williams
73 Georgina D. Youngs
74 D.A. Baster
75 Brian A. Southern
76 Stephen John Green
77 Miss M.P. MacFaul
78 Mr & Mrs R.T. Wilkinson
79 Dr C.I. Morgan
80–81 Mr & Mrs R.H.C. Stewart
82 B.J. Clem
83 David J. Warner
84 Frederick James McCanch
85 Norman Victor McCanch
86 Miss B.C. Hibberd
87 G.H. Knight
88 David Stacey
89 Richard Darwin
90 Mrs. J. Morgan
91 Miss G.E. Morris
92 S. Sutcliffe
93–95 Betty Crockett
96 K.J. Thomas
97 Mrs A.J. Hugill
98 John Rees
99 The Woodland Trust
100 Dr J.M. Stuart
101–102 Vivienne Scale
103 G. Smalley
104 Geoff Battershall
105 Dilwyn Roberts
106 R. Hartnup
107 R.I. Millichamp
108 Mrs R.F. Maud
109 J.W. Thomas
110 Mrs A.A. Lloyd-Williams
111 W. Bowman
112 S. Tiller
113–114 Mrs M.B. Thomas
116 L.A. Cram
117 A.R. Lee
118 Mr & Mrs J. & A. Poole
119 Miss E.S. Mathias
120 A.R. Lee
121 J.B. Keatley
122 M.P. Swire
123 R.S. Tubbs
124 H.T. Conway
125 G.C. Lambourne
126 John Comont
127 Lawrence E. Wade
128 D.M. Allen
129–130 James Ogden
131 B.W. Stumpe
132 R.A. Kennedy
133 Mrs J. Ford
134 Forestry Commission
135 E.C.K. Redd
136 Margaret Hughes
137 William David Llewelyn
138 Iris Silvester
139 G. Harrison
140 C. Winnard
141 Helen Whalley
142 J.R. Mathias
143 Miss M.E. Acors
144 D.L. Booth
145 Alan Maddock
146–147 John Richard Ellis
148 Anne M. McCall
149 G.R. Tait
150 C.M. Stenger
151 Ewan Thomas
152 Mrs G. Boote
153 John Harvey
154 Gordon Smalley
155 Jonathan Stanley Aylett
156 M. Deighan
157 Barbara Last
158 R.C. Tattersall
159 Mrs J. Kimpton
160 Natasha de Chroustchoff
161 R.B. Wright
162 Anthony David
163 L.W. O'Rourke
164 R.A. Spencer
165 R.C.O. Davies
166 Andrew R. Hill
167–168 M.B. Jeeves
169 David G. Rotter
170–171 Peter Bossom
172 Dr A.K. Jones
173 Kathrin Rose
174 Bernard & Dulcie Coldicott
175 Mrs E. Evans
176 J.R. Cole
177 A.E. Watts
178 W.A. Emerson

179 W.H. Brick
180 I.W. Young
181 M.W. Newton
182 Dr & Mrs John Cule
183
184 Dr John Herbert
185 John Easton
186 Gwyneth Morris
187 Joint Library, Welsh Agricultural College
188 Prof R.G. Newton
189 Stephen Grahame Gill
190 David John Jones
191 Paul M. Burnham
192 C.M. Pryor
193 Mrs B. Pitter
194 Mrs B.M. Sandy
195 M.R. Lippiatt
196 G.C. Davis
197 R. Jennings
198 Ian Scott
199 R.W. Cunningham
200 Joyce Gwynne
201 L. Taconis
202 Edward Bolger
203 Dr A.D.Q. Agnew
204 John Hugill
205 John Stott
206 D.T. Lewis
207 Nigel Hobday
208 Thomas Latter
209 Lt Col G.G. Roach
210 John W. Evans
211 K.W. Longstaff
212 Daphne Shrager
213 R.E. Hewitt
214 Bernard J. Wright
215 Margaret Patterson
216 Hefin Jones
217 Miss I.M.N. Ryan
218 F.B. Cook
219 Joyce M. Harris
220 F.H. Alderson
221 Jonathan Alan King
222 Mrs Llinos Taylor
223 S.C. Harrison
224 Mrs M. Whittaker
225 W.H.C. Williams
226 Mrs E. Mellor
227 P.J. Whitcomb
228
229 Mrs J.P. Savidge
230 David John Jones
231 Peter Brown
232 R.J. Wootton
233 G.W. Rainey
234
235 Doreen Cooke
236 Anthony David Jones
237 David Harries
238 Miss Kay Skippins
239 Miss B.T. Clay
240 Trevor Theobald
241 T.C.E. Hughes
242 Derek & Claire Bayliss
243 L.R. England
244
245 C.A. Betts
246 Duncan S. Collins
247 A. Ashman
248 Mrs M.J. Rogers
249 B. Rosenfeld
250 W.A. Croft
251 R.M. Roper
252 W.S. Duckering
253 J.R. Oliver
254 K.J. Doughty
255 H.F. Lamb
256 M.J. Dunn
257 Olwen Deibert
258
259 Quarry Trading Co Ltd
260 R. Manson
261 C.M. Batt
262 D.R. Rees
263 A.E. Bevens
264 Elizabeth Ruth Davies
265 C.W.C. Dawkins
266 M.I. Evans
267 Mrs Winifred Bowman
268 A.O. Chater
269 R. Hugh A. Flack
270 Mrs J. Ball
271 D. George
272 Miss R.P. Setchfield
273 Mrs Susan Hewitt
274 Mrs Carolyn Hunt
275 Miss D.A. Phillips
276
277 Robert Saunders
278 Mrs Hilda Fenner Clayton
279 A.J. Hansen
280 Charles Mark Laird
281 Mrs R.A. Coldwell
282 G.C. Lambourne
283 Dr L. Gibbons
284
285 Mrs N.J. Cox
286 Paul Jackson
287 J.J.S.V. Lloyd-Williams
288 J.W.A. Lloyd-Williams
289 D.W. Benjamin
290 J. Green
291 John Valentine
292 Ian S. Watt
293 William Condry
294 Geoff Liles
295 Joan Harlan
296 A.K. Pearce
297 Mrs D. Rees
298
302 Haverfordwest Library
303 Margaret Crouch
304 Adrian Hughes
305 B.J. Reeley
306 G.H. Green
307 J.M. Jones
308 The Rt Hon John Wakeman MP
309 Ann Creswell
310
311 R. Dynevor
312 Helen Davidson
313
314 Shelagh Seery
315 V.M. Williams
316 Carmarthen Area
321 Library
322 Dr G. Hutchinson
323 Rev Everard Williams
324 D.E. Davies
325 Mair Shanks
326 Catherine Lloyd
327 Mrs Joy Lee
328 Martin Andrews
329 Michael Worley
330 R.F. Ashton
331 Mrs Elaine May
332 Mrs Margaret Chater
333 E.R. Carr
334 N.N. Beach
335 Ruth Griffiths
336 Dorothy J. Herlihy MBOU
337 Mrs J.L. Dally
338 K.W. Lynch
339 John Coales
340 Leonita Slay
341 R.V. Wilks
342 M. Bellamy
343 Jean M. Hemming BSC
344 M.P. Ireland
345 J.. Havard
346 Prof W.E. Waters
347
348 Eugene Budgen
349 W.L. Rudland
350 Valerie Jones
351 Edward Arney
352 E.C. Deadman
353 Dr Richard B.L.
355 Edwards
356 Michael D. Lort
357 J.P. Robinson
358 Mrs M. Williams
359 A. Bonham-Noyle
360
361 Elinor Gwynn
362 Mrs G.M. Heal
363 Huw Edwards
364 G.R. Shannon
365 Dr Huw Edwards
366 R.A. Harper
367 Professor R.J. Berry
368 John L. Oakley
369 M. Quartermaine
370 Peter Conder
371 A.P. Fowles
372 R.J. Latham
373 Mrs Irene Weston
374 Eric Ayers
375 A. Diamant
376 T.G. Martin
377 Mrs H. Cowman

378 Mrs E. Foster
379 R. Morgan
380 Dr M. Silverman
381 D.J. Smith
382 C.H. & J.E. Paris
383 Denise Long
384 M.R.L. Johnston
385 Trevor J. Price
386 David Russell Barnes
387 David Reaney
388
389 Ivor Griffiths
390 Mrs Blatchley
391 Q.O.N. Kay
392 David Henry Davies
393 Dr Maldwyn Thomas
394 Dr Noel Thomas
395 Veronica Morris
396 Prof M.F. Claridge
397 Chapter & Verse Bookshop Ltd
398 Martin Palmer
399 Ivan & Catherine Vawdrey
400 Mrs Elaine Hemming
401 T.C.E. Hughes
402 D.K. Thomas
403 Mrs K.R. Evans
404 G. McLardy
405 Michael McLardy
406 P.N. Ferns
407 D.S. Meyler
408 Miss J. Paterson
409 Rosemary Leach
410 David Hedley Williams
411 Mrs A. Poole
412 Eric Bray
413 Peggy Cateaux
414 Miss H.J. Tasker
415 L.G. Broomfield
416 Mrs C. Griffiths
417 E. Barnes
418 Thomas Linfoot
419 Anne M. Bryan
420 David J. Donati
421 B.J. Gregory
422 R.P. Bray
423 M.A.R. & Mrs C. Kitchen
424 B.T. Lewis
425 W.A. Thomas
426 David Robert Davies
427 Mrs L. Morris
428 David Green
429 Thomas William Davies
430 Prof R.J. Berry
431 E.F. Evans
432 Emeritus Professor William S. Lacey
433 Anne Robinson
434 West Glamorgan
437 County Library
438 S.J. Preece
439
440 Mrs Barbara J. Jones
441 L.P. Thomas
442 J.R. Cole
443 Alan Orange
444 John R. Haynes
445 Dr Douglas B. Kell
446 Miss O.E.G. Evans
447 Ieuan Williams
448 H.W. Roderick
449 Paul Jackson
450 T. Goodwin
451 Dr W.M. Kurowski
452 Dr Keith Gull
453 M.J. Coveney
454 Francis Bunker
455 Richard Perry
456 Hillscape Walking Holidays
457 Valerie Linda Dixon
458 P.H. Ross
459 Elizabeth Jones
460 John Meredith
461 F.M. Slater
462 E. Andrews
463 Z.M. Owens
464 P.J. Dobbs
465 Peter Francis Coveney
466 R.G. Greenslade
467 Major Vince Gwilym RAMC (V)
468 M.G. Evans
469 Veronica Shrubb
470 Jeremy H. Williams
471 Dr Alan Morton
472 Martin Peers
473 Dr M. Gillham
474 Miss Betty Crawford
475 K.E. Stott
476 Raymond Phillips
477 Keith Alexander
478 G.J. Morris
479 P.D. Dickin
480 R. Liford
481 Grace & Eric Nicholson
482 David Davies
483 John Vivian Hughes
484 L.E. Williams
485 Christopher West
486 Mary Scruby
487
488 T.J. Heal
489 Robin Bunnage
490 D.A. Bullen
491 Dr J.R.S. Webb
492 The University College of Wales, Acquisitions Librarian
493 Paul Kempster
494 Richard Howell
495 Dr G. Hutchinson
496 C.L. Perry
497 Mrs Sylvia R. Mitchell
498
499 Raymond Garlick
500 J. Cecil Goodwin
501 Susan Maling
502 Paul G. Billington
503
504 R.W. Hobbins
505 G.C. Lambourne
506 Andrew & Margaret Silander
507 E.W. Bence
508 Catherine Jane Howarth
509 G.A.K. Brunt
510 J.A. & J.D. Stephenson
511 Bryan Watkiss
512 Ton Wildschut
513 Jane E. Hodges
514 John Newton
515 T.E. Davies
516 Frederick Irwin
517 E.E. Owen
518
519 J.M. Hedger
520 E.M. Horley
521 J. Justin Yates
522 A.R. Henman
523 Martin Ware
524 Dr Robin G. Crump
525 Vera K. Hughes
526 Alex & Joanne Maltham
527 Miss M. Buze
528 Miss D.E.G. Evans
529 S. Truelove
530 Dr W.J. Jordan
531 Gwenda Meinir Jones
532
533 Coleen Thomas
534 Huw Lewis Jones
535 Graham Donald Arnold
536 Wyn Lloyd Parry
537 J. Gordon James
538 C.J. Mansfield
539 Mrs E.M. Laing
540
541 Miss June Morgan
542 R.W. Cooke
543 E. Lawrence Bee
544 Olive Winfield
545 Sir Edward & Lady Parkes
546 Peter & Stephanie Thomson
547 Peter Williams
548 Ian & Jennifer Stewart
549 Martin & Julie Jones
550 Betty Carey
551
552 L.J. Phelps
553 Mr & Mrs J. Johns
554 Mr & Mrs R.W. Clutterback
555 Mrs Peggy House
556 Mr & Mrs W.E. Rooker
557 G.W.F. Lloyd
558 Miss V.A. Davies
559 P.J.M. Nethercott
560 W.M. Condry
561 Nigel Fellows

562 Jonathan W. Adams
563 Jeremy Moore
564 R.M. Badcock
565 Neil E. Owen
566 Patricia Treves
567 K.R. Ineson
568 R.T. Jones
569 John Dietz
570 Helen Strojek
571 S. Davey
572 Mrs J.M. Andow
573 M.W.D. Brace
574 J.A. Woodley
575 Mr & Mrs P. Sharpe Neal
576 Mr & Mrs B. Parks
577 Anthony J. Barbasio
578 S.V. Wolfe
579 Geoffrey H. Battershall
580 Goronwy Wynne
581 Elizabeth M. Watson
582 Prof J. Green
583 Lawrence Rawsthorne
584 C.R. Bantock
585 Derek Bartley
586 Ewart Hughes
587 Dorothy E. Cornwall
588 Martin Humphreys
589 David J.P. Miller
590 Mrs M.J. Morgan
591 Dr C. Lloyd-Morgan
592 Mrs L.A. Cadbury
593 Dr Louise F.W. Eickhoff
594 A.G. Lacey
595 Vernon Hughes
596 Jonathan E. Marsh
597 Graeme M. Kay
598 Twn Elias
599 Dr D.A. Dorsett
600 David Egan
601 J.T. Hagarty
602 M.W. Kilby
603 W.K. Rose
604 S.B. Reynolds
605 H. Clutton-Brook
606 R. Jean Conran
607 Mrs D.M. Owen
608 David J. Cooksey
609 Davina Rendall
610 C.K. Britton
611 Margaret J. Lockwood
612 R.W. Parr
613 Mrs E.W. Weyman
614 Freda Williams
615 Marjorie Stoddard
616 T.R. Phillips
617 C.S. Briggs
618 Miss E.J. Allen-Williams
619 Mrs D.R. Pownall
620 Peter Schofield
621 Pat Denne
622 Robert Edwards
623 Christopher M. Swaine
624 Brian A. Southern
625 E.G. Reynolds
626 H.T. Conway
627 P.W. Swire
628 Miss E.S. Mathias
629 J. & A. Poole
630 West Wales Trust for
679 Nature Conservation
679 Mary Thomas
680 Andrew Robert Lee
681 Andrew R. Lee
682 R.I. Millichamp
683 R. Hartnup
684 C.K. Britton
685 Geoff Battershall
686 J.B. Keatleyt
687 G. Smalley
688
689 Vivienne Scale
690 Dr. J.M. Stuart
691 Joyce & John Rees
692
694 Betty Crockett
695 Miss G.E. Morris
696 David Stacey
697 C. Fuller
698 W.H. Gravell
699 Dr C.I. Morgan
700 R.T. Wilkinson
701 W.K. Rose
702 R. Jean Conran
703 H. Clutton-Brock
704 I.W.R. Spriggs
705 Brian A. Southern
706 Dr & Mrs W.G.G. Loyn
707 Mary P. MacFaul
708
709 R.D. Pryce
710 I.K. Morgan
711 R.E. Saunders
712 M.E. Tiller
713 D.V. Newman
714 Mrs M.D. Ravenswood
715 Joseph Lewis
716 A.P.R. East
717 Miss Margaret Batt
718 Fiona Benn
719 Sheridan Newton
720 S.J. Sutcliffe
721 David Baster
722 Pembs Coast National Park
723 Woodland Trust
724 Winifred Bowman
725 Isobel & Eric Soper
726 Ralph S. Tubbs
727 Huw Phillips
728 Mrs A.A. Lloyd-Williams
729 John P. Palmer
730 Mark Johnson
731 H.F. Lamb
732 John S. Barker
733 David Hearnshaw
734 Omar Yasseen
735 Catherine Saunders
736 Rachel Saunders
737 B. Powell
738 Wynne J. Samuel
739 R.H. Craythorne
740
741 M. Panter
742 Robert Lewis
743 H. Thomas
744 Llyfrgell Genedlaethol Cymru — The National Library of Wales
745 G.D. Dobbins
746 Peter Coath
747 J.E. Owen
748
749 Mrs M. Briers
750 Monty Philpin
751 Ken & Cleone Gardner
752 John & Ann Allen
753 D.C. Davies
754 E.B. Jones
755
756 Anne Thomas
757 Mick Green
758 E.J. Thomas
759 W.E. Davies
760 M. Glossop
761 Sister Dolores
762 C.M. Batt
763 Peter Hope Jones
764 Ann Conolly
765 Jeffrey Taylor
766
767 Harold Platt
768 Robert Jones
769
770 David M. Jones
771 Margaret R. Hodson
772 Reginald A. Long
773 Mr & Mrs R. Durbin
774 N.P. Allen
775 Mr & Mrs N. Casingena
776 Nichola Wilkins
777 Sian Miller

Remaining names unlisted.

ENDPAPERS: Nature Conservation sites in Dyfed.

Visitors landing on Skomer. The provision of facilities for people to enjoy and understand the wildlife of West Wales is a key aspect in ensuring its survival. (NCC)

Nature Conservation Sites in DYFED

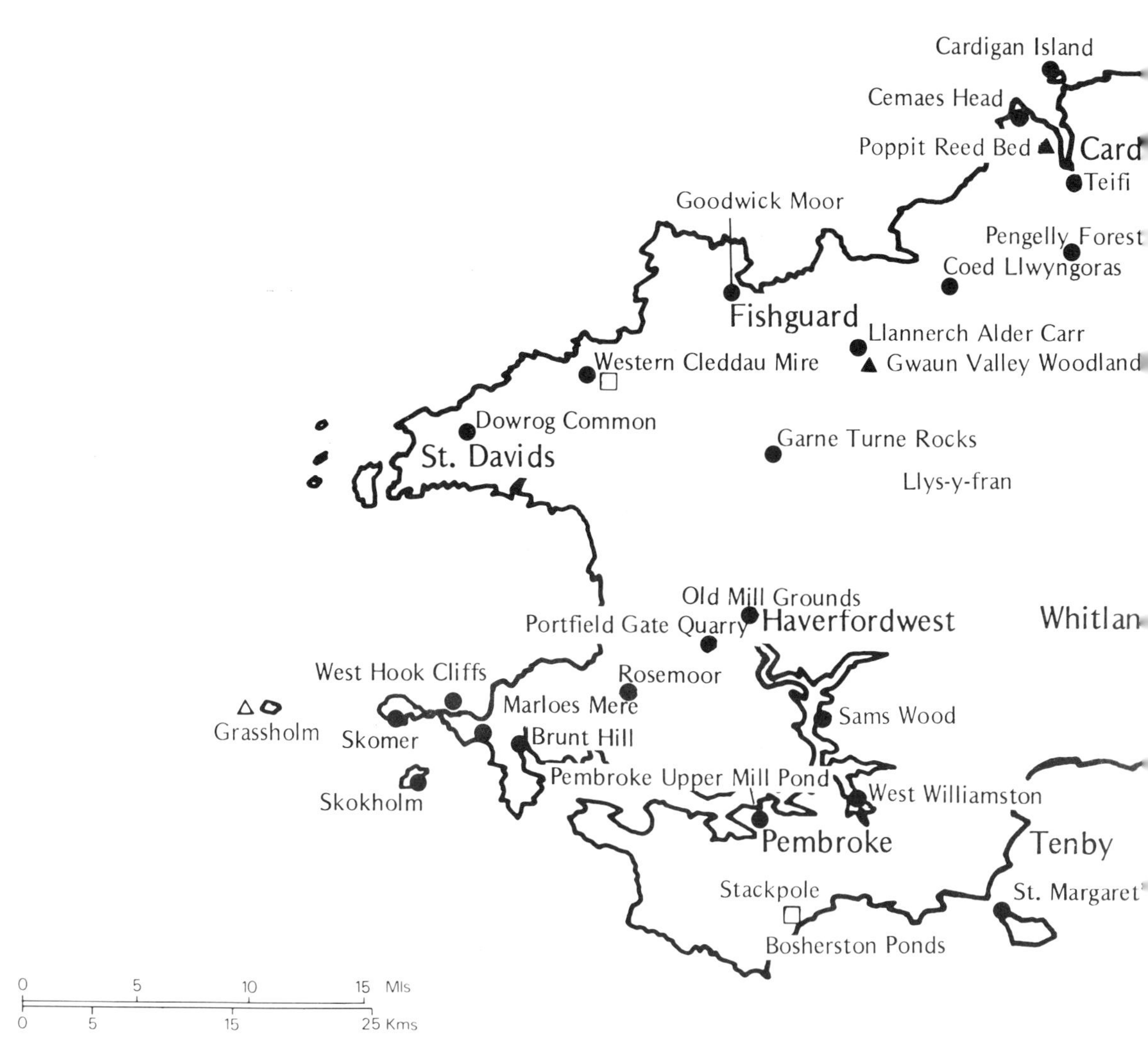